Astronomical Symbols
on Ancient and Medieval Coins

To my wife, Alice, for her ongoing encouragement to complete this book.

To my sons, Michael, Jim, and Dan, for giving up time that would have been spent with Dad during the writing of earlier versions of this book.

To my late dog, Andy, for the countless hours he lay at my feet while I worked at the computer, hoping that I would reduce the number of times I stepped on his nose.

To the late Dr. Stanley P. Wyatt, author and astronomy thesis advisor at the University of Illinois at Urbana-Champaign, for instilling within me a greater appreciation for the wonders of the solar system.

To the late E.R. Duncan Elias, for his excellent book on Anglo-Gallic coins, in which his detailed information started my pattern match between astronomical symbols and actual celestial events, and for his desire to meticulously describe the details and varieties of the most common coins.

Astronomical Symbols on Ancient and Medieval Coins

Marshall Faintich

McFarland & Company, Inc., Publishers
Jefferson, North Carolina, and London

The present work is a reprint of the illustrated case bound edition of Astronomical Symbols on Ancient and Medieval Coins, *first published in 2008 by McFarland.*

LIBRARY OF CONGRESS CATALOGUING-IN-PUBLICATION DATA

Faintich, Marshall, 1947–
Astronomical symbols on ancient and medieval coins / Marshall Faintich.
p. cm.
Includes bibliographical references and index.

ISBN 978-0-7864-6915-4
softcover : acid free paper ∞

1. Astronomy in numismatics. 2. Numismatics, Ancient.
3. Numismatics, Medieval. 4. Coins, Ancient. 5. Coins, Medieval.
6. Astronomy — History — Miscellanea. I. Title.
CJ161.A82F35   2012   737.49'3 — dc22   2007020104

British Library cataloguing data are available

© 2008 Marshall Faintich. All rights reserved

No part of this book may be reproduced or transmitted in any form or by any means, electronic or mechanical, including photocopying or recording, or by any information storage and retrieval system, without permission in writing from the publisher.

On the cover: background © 2012 Shutterstock.
Coins clockwise from top: Denarius reverse; Athenian tetradrachm reverse; Athenian tetradrachm obverse; Stephen mullet coin reverse and obverse (from the author's collection)

Manufactured in the United States of America

*McFarland & Company, Inc., Publishers
Box 611, Jefferson, North Carolina 28640
www.mcfarlandpub.com*

Table of Contents

Acknowledgments — vi
Preface — 1
Introduction — 3

1. The Sun and the Moon — 9
2. The Stars and the Planets — 17
3. Eclipses — 32
4. Comets — 59
5. Complex and Unusual Astronomical Designs — 82
6. Visions Explained — 98
7. Norman England — 122
8. Re-attribution of Anglo-Gallic Deniers — 153
9. The Solar Crusade — 170
10. Beginnings — 178

Appendix A. Accuracy of Analyses — 181
Appendix B. Additional Examples — 189
Appendix C. Sources of Figures — 214
Bibliography — 217
Index — 221

ACKNOWLEDGMENTS

This volume does not stand alone. It is the result of the analysis of historical and numismatic data that have been compiled for almost 2,000 years. From the ancient chronicler to the 19th and 20th century numismatic scholars, I am indebted.

Some of the coins depicted in this manuscript are from my private collection. Many more coins are difficult to obtain, and, in some cases, are unique. Therefore a few of the figures have been extracted from other books, and a large number of photographs of coins that served as the source for my drawings came from a multitude of books, catalogs, and auction and sales materials. Sources for the extracted figures are listed in Appendix C, and I am grateful for the permission received to use figures that were copied from other sources.

I also wish to thank the many readers of my earlier works on this subject for their reviews and helpful comments. Special thanks for their excellent numismatic and historical comments go to Bruce Brace, president of the Classical and Medieval Numismatic Society and the honorary curator of the numismatic collection at the McMaster Museum of Art, and to Owen Gingerich, senior astronomer emeritus at the Smithsonian Astrophysical Observatory and research professor of astronomy and of the history of science at Harvard University.

Two other reviewers deserve special recognition. Victor Failmezger, author of *Roman Bronze Coins*, provided detailed comments and expert opinions on the historical and numismatic aspects of Roman coinage discussed in this book and made greatly appreciated suggestions on the form and content of this manuscript. Steve Ford, one of the world's experts on Anglo-Gallic coinage, co-authored a portion of chapter 8 and provided detailed comments on the rest of the chapter.

Any errors or omissions in this book are my own. Even my expert reviewers did not always agree with each other, but such disagreements were usually on topics where no clear answer is known.

And finally a special thank you to my wife, Alice, a professional editor whose talents are in great demand, and whose time spent editing this book added many hours to her already busy days. If any editorial or style errors remain in this book, they must be attributed to my failing to make changes per her comments.

Preface

Ever since the first coin was struck nearly 3,000 years ago, sovereigns realized that these small bits of metal were a powerful medium for conveying propaganda to the general populace. Battle victories, religious deities, and other symbols were placed on coins to show not only the strength of the reigning sovereign, but also a divine right to the throne. Astronomical symbols represented pagan gods, and such symbols were widely used as well. When a spectacular celestial event occurred, such as a bright comet, an eclipse, or a tight clustering of planets, sovereigns were quick to capitalize on what the populace had seen in the heavens.

As ancient civilizations in Europe and nearby lands moved from pagan beliefs to Christianity, the symbols also changed from showing celestial gods to Christian representations. However, astronomical symbols portraying astrological events continued to be struck on coins through the end of the medieval era. No matter how much the church tried to stop astrological beliefs, the heavens remained a powerful force on Earth. Even though astrology was in direct conflict with Christianity, using celestial symbols as propaganda to demonstrate a divine right to authority was too valuable of a tool to be discarded.

Writing this book has been a passion for me. Ancient and medieval numismatics, history, cartography, and astronomy are the foundations of this book. Techniques in spatial and statistical data analysis, pattern recognition, and correlation of diverse types of information have been my tools. The combination of these disciplines was required to make the first pattern match that started my research on this topic. When I saw a figure of the 1290 denier of Edward II of Ponthieu, which depicts a crescent, an annulet, and then another crescent, my immediate thought was that it represented the phases of an annular eclipse. When I discovered that an annular eclipse actually did cross Ponthieu in 1290, I was hooked. From that moment on, the pieces of the puzzle began to come together, one coin at a time.

My first attempts to put my research into words resulted in various versions of a self-published book, *Symbolic Messengers of Medieval Man*. Originally published in 1993, with three major revisions during the next two-and-a-half years, this book focused on the use of astronomical symbols on medieval coins and only touched on the ancient coins. Unfortunately, I had some erroneous examples in the earlier versions, but thanks to the many welcomed comments from more than 200 readers, I eventually weeded out most of the offenders.

Part of my original research included the Roman coinage of Constantine the Great, and this led to my writing a second book, *Visions Explained*, that was completed in 1998. This book was much too short to stand on its own, and after making 20 or so copies for interested readers, I decided to completely rewrite my first book and include an expanded version of *Visions Explained* as a chapter.

Some readers of this book may attribute the matches between astronomical design symbols on ancient and medieval coinage and actual celestial events to be merely coincidental, or perhaps a forcing of the data to fit the events. Any poor examples would add credence to arguments that oppose the hypotheses put forward in this volume; however, so many coins are presented here that support the hypothesis that a few incorrect examples must be considered as noise in the analyses.

As I began to write this book, I thought I was writing a book about numismatics. By the time I finished, I had found that the book is much more than that. It is the basis for a new tool to explore and understand the political and mystical beliefs of ancient and medieval life. Thus I have not finished a book, but rather I have opened the door for more research. This is the passion: each new discovery answers one question, and at the same time asks another. Where will it lead? I do not know, but I invite you to join me on the journey.

Marshall Faintich
Spring 2007

INTRODUCTION

The change of a kingdom was determined by the heavens. In this illustration, a physician and an astrologer attend the death of William II of Sicily in 1189. This figure is taken from De rebus siculis carmen, *a circa 1196 poem by Pietro d'Eboli, courtier to the king (Lopez 1966).*

Battle victories, great fortresses, and any other propaganda that sovereigns could use to support and preserve their power found their way into the designs of ancient and medieval coins. Everyone from local merchants to soldiers might have one or more coins in their possession, and those little pieces of metal were extremely important to the owners. Coins were handled carefully, evaluated for metal fineness and weight, and inspected for genuineness. What a perfect medium for rulers to convey political messages to their subjects.

Rulers used depictions of deities and celestial events on their coinage as propaganda devices to reflect their divine right to the throne. The heavens were not only a mystical force to ancient and medieval peoples, but a foreteller of the future, while man simply played out the story according to a predetermined plot.

The use of deities and celestial events on coinage to reflect this divine right is found on the earliest ancient coins through those of the Middle Ages. One of the primary symbolic devices to portray this concept was the "hand of God" reaching down from the heavens (figure 1).

Figure 1. England: Aethelred II (978–1016 A.D.), showing the "hand of God" on the reverse

When an extraordinary omen occurred, such as an unusual celestial event, the sovereign was twice blessed. Comets, eclipses, and close gatherings of planets explained the present and declared what would happen in the future. Everyone from the smallest child to the opposing warrior could witness the event, and rulers were quick to order their die engravers to capitalize on these heavenly sources of propaganda.

Ancient cultures mixed pagan religious beliefs with scientific measurements of the heavens. As Christianity spread across Europe during the early medieval period, ancient knowledge was lost or destroyed. Although translations of Arabic knowledge into Latin began in the second half of the 11th century (primarily in Muslim Spain and Sicily), it was not until late in the 14th century that the astronomical knowledge of the Arab and ancient Greeks was fully discovered and translated. Until that time, astrology and astronomy were interchangeable concepts, surrounded by myths and superstitions.

People's view of the natural and the supernatural was reversed from that of modern times. They saw the supernatural as the natural order of events and accepted its phenomena as valid, while at the same time they dismissed any evidence offered by the material world. Stars were signs of physical processes and events, such as sickness and weather. Eclipses and comets were associated with storms, floods, droughts, and earthquakes. Comets, especially those with long tails, were also seen as portents of personal and general disasters, wars, changing kingdoms, and deaths of princes. Comets were classified as belonging to different planets and shared the superstitions surrounding their planetary counterparts.

Saturn presided over life, science, and buildings, while Jupiter controlled wishes, honors, riches, and garments. Mars was associated with wars, individuals, marriages, and feuds. Venus influenced love and friendship, and Mercury controlled disease, debt, fear, and commerce. Hope, happiness, and gain came from the sun, and the moon caused wounds, robberies, and dreams. Without a comet to blame, astrologers attributed the outbreak of the black death in the 14th century to the conjunction of Saturn, Jupiter, and Mars.

The most significant celestial events are eclipses of the sun and moon, conjunctions (close groupings) of the planets, and spectacular comets. Given the rarity and awesomeness of such events, coupled with the strong religious beliefs and astrological mysticism of ancient and medieval times, many people interpreted them as signs or omens.

Celestial events played a major role in everyday ancient and medieval life. Contemporary historians recorded battles, droughts, floods, plagues, famines, eclipses, comets, and the death and accession of rulers and religious leaders as the greater part of their subject matter. Some of the more spectacular celestial events were symbolized on contemporary coinage.

Acceptance of coinage was necessary for a region's economic stability, and coinage designs quite often became immobilized (that is, remained the same), or underwent only minor changes over long periods of time. Thus a particular design feature may represent an astronomical event that occurred many years before the coin was issued if it also appeared on earlier coins of that type. Therefore it is important to determine when a design change or the addition of astronomical symbols first occurred for the type of coin under consideration.

Minor design elements, including astronomical symbols, were sometimes used to designate a moneyer, mint, or minor change in coin weight or content. One can argue that astronomical symbols used for this purpose have no relationship to specific celestial events; however, actual celestial events may have provided the inspiration for such marks.

Some numismatists attribute any correlation of astronomical events with coin design to pure coincidence. Thus four factors must be present to substantiate the correlation as non-coincidental. First, the date of a coin bearing an astronomical symbol must be ascertained. Second, the astronomical symbol must be the first such occurrence for that coin design or a re-introduction of the symbol after a substantial period of time to rule out immobilization of the design. Third, the occurrence of the astronomical event must be established. Fourth, and most difficult to ascertain, historical evidence must be presented that supports the observance and importance of the event. Without the latter, any correlation between a design symbol and an astronomical event is merely speculation.

The first three tasks can be accomplished with only modest difficulty by examining modern numismatic collections and recreating the celestial heavens for any date in history. Many years of numismatic research have resulted in a fairly good assessment of coinage issue dates and design sequences, and most types of astronomical events can be re-created with reliable precision.

The observation and importance of celestial events may be determined even if the coins themselves are the only source of historical evidence. Sometimes a coin with astronomical symbols was issued where the celestial event was not recorded; however, in some of these cases chroniclers in other regions recorded the event, implying that the event may also have been visible in the region where the coin was struck. The addition of the same or similar astronomical symbols to coins different rulers struck during the same period may indicate the importance they associated with the celestial event.

Even with the presence of all four factors needed to substantiate the correlation between an actual celestial event and an astronomical symbol on a coin, no one today can state with certainty what was in the mind of the ancient or medieval engraver who prepared the dies for the coin. However, with enough examples of possible correlations between actual celestial events and symbols on many coins, a strong argument can be made that using these events and their symbols as propagandistic devices was common practice.

Examining the Evidence

The coins of ancient and medieval Europe offer a unique insight into the economic and political history of the times. Certainly the designs on coinage reflect the life, propaganda, and beliefs of these eras. Many of the symbols incorporated into coinage design have been treated by scholars simply as artistic forms, when in reality the symbolic messages of the designs were a basic form of communication in a world where few could read or write, but all could observe the heavens and see similar shapes on everyday coinage.

In addition to military, civil, and religious motifs, ancient and medieval coins are also decorated with symbols such as pellets (dots), annulets (circles), stars, crescents, mullets (stars with a hole in the center), and other celestial markings. The interpretation of these symbols, along with their relationship to actual physical events, provides a unique opportunity to better understand the underlying political conditions, astrological beliefs, and corresponding dates related to the issuance of these coins. Yet for hundreds of years, astronomical symbols on coins remained overlooked as a chronicle of the history of ancient and medieval thought. The recent development of computers and sophisticated mathematical models of the solar system set the stage for multidisciplinary research involving astronomy, history, and numismatics.

Re-creating a historical event requires assembling a variety of sources of information. Each piece of evidence must be reviewed with some skepticism, as some sources will be more reliable than others. Clearly when a divine vision or omen is part of the puzzle, the task at hand is to study all the physical evidence and then try to surmise what was in the minds of the participants. While difficult, this is not a totally impossible venture. Indeed, with enough circumstantial evidence, one can make a fairly strong case as to what may have occurred.

The fact that historical written accounts of ancient and medieval happenings are among the least reliable sources of information may initially seem strange, but consider the following: all chroniclers presented their own point of view of history, thus their accounts may have been biased by personal, religious, and political forces. Remember that few people at the time had access to the written record, and chroniclers surely did not go out of their way to get critical reviews from those with opposing viewpoints.

Another problem is translation from one language to another. More important, as manuscripts were copied and re-copied by hand, each copy was subject to additional re-interpretation errors and blunders. Finally, some accounts were written soon after an event took place, while others were recorded years later, allowing plenty of time for embellishments and changes to have entered the spoken tale. Nevertheless, the written record is a good starting point, and the greater the number of written sources describing a certain event, the easier it is to weigh one version against another.

Historical records of celestial events provide another source of information to piece the puzzle together. Celestial events were often better recorded in eastern chronicles than in European writings. Modern computers and astronomical software programs permit us to verify and complete the historical record by re-creating the heavens at any time in history, and can be programmed to find certain types of events that may have been witnessed. However, such re-creations do not guarantee that the celestial event was observed where a coin was struck with an astronomical symbol that seems to relate to that event.

Fortunately, other sources of evidence are available. Paintings, pottery, tapestry, and other forms of ancient and medieval artwork often recorded historical events. Although these media still contained biases, they were subject to visual scrutiny and did not require that reviewers had the ability to read written text. However, the distribution of most ancient artworks was extremely limited, and thus few people were able to verify their accuracy. By contrast, the one form of artwork that usually can be relied upon is ancient and medieval coinage. Just as celestial events made their mark on the painting, tapestry, and literature, they were also depicted on coinage. While coinage design was subject to the same biases as the written record, multiple copies of coins were produced and were distributed to a wide sector of the population. Sovereigns could not allow the truth to vary too much from reality or from what was believed to have happened without losing credibility with their subjects. In addition, large numbers of metal coins have survived while much of the written medium has been lost.

Quite often, coins are the only man-made source of information about a period of time. Ancient engravers of coin dies were often careful to picture their rulers as accurately as possible, unless they were trying to improve on their appearance. When sovereigns were in power for many years, coin portraits that spanned their reigns have been used to determine the aging, and sometimes the medical conditions, of the sovereigns. During the medieval period, some historical evidence was lost as the artistic excellence of ancient die engravers was replaced by crude depictions.

Ancient and medieval coins do not show the date that the coin was issued, although some coins record the regnal year of the depicted sovereign; however, research that includes examining other objects found with coin hoards; ordering coins temporally by means of changes in their size, weight, and metallic fineness; and linking of dies used to produce coins from one issue to the next serves to establish relative dates of issuance. With the addition of coins commemorating historical events that can be dated, dates for entire series of coins can be established with a high degree of certainty. While some coins can be dated only to within a half century of being issued, others can be dated to within a year or so.

Correlation or Coincidence?

The coinage of ancient and medieval Europe is abundant with astronomical symbols. Interestingly, as the need for larger silver coins and gold coins increased during the later medieval period, astronomical symbols tended to remain on the common or smaller coins and on those with less silver content. These were the coins that were the mainstay of everyday business, rather than the riches hoarded by the nobility. Perhaps astronomical symbols were used as a propaganda device to influence the support of the common people by convincing them that the accession of new rulers, the demise of old rulers, and battlefield victories were divine acts.

The concepts presented here are not entirely new. For more than a century, a handful of other researchers has investigated the significance of astronomical symbols on ancient and medieval coinage, yet the correlation between such symbols and actual celestial events has remained controversial. Current numismatic thinking clearly attributes stars and crescents to the sun and the moon, but offers little additional interpretation.

Certainly the possibility exists that pellets, annulets, and the like were sometimes used only for design purposes, and at other times to signify something other than an astronomical event. Some may dismiss the entire hypothesis of this book as pure coincidence, while others may find individual examples that do not fit the hypothesis, and therefore deny any correlation whatsoever, but the evidence for correlation is too strong.

The following chapters will build the case for the hypothesis that astronomical symbols on coins often represent actual celestial events. Examples are too numerous for all of them to be discussed, but the reader shall travel from one end of Europe to the other, and shall see a consistent pattern that transcends 20 centuries as well as diverse cultures. This volume will show that the possibility of the use of astronomical symbols on coinage as commemorating actual physical events is too consistent to be pure chance. Once the correlation is accepted, the dates of the celestial events can supplement the historical record to help answer questions about political, economic, and religious history.

THE SUN AND THE MOON

Crescent moon above the owl of Athena (c. 490 B.C.) that was taken as a victory omen during the Battle of Marathon

Astronomical symbols have been depicted on coinage almost since the first known use of coins. Whether early designs were used to depict actual celestial events, deities, or generic heavenly bodies is difficult to ascertain. What is clear, however, is that astronomical symbols were used on a great variety of coins of many different lands from ancient cultures though the medieval period.

The oldest series of known coins (figure 2) was struck around 700 B.C. in Ionia, now part of western Turkey. Prior to this series, precious metals were used for trade, but each piece had to be weighed individually to determine its value. This first series of coins, a stater and fractional pieces, formed the basis for a set of standard weights for trade. Electrum, an amber-colored alloy of gold and silver, was used for the first coins. Interestingly, one form of the Greek word for electrum translates as "the beaming sun."

Figure 2. Ionia: ½₁₂ stater

Striation marks are found on one side of the coin, and one or more punch marks were made on the reverse to permit verification of the quality of the metal. A sharp blow from a hammer onto the punch made the mark on the reverse side of a blank metal globule, and simultaneously, the metal was forced into an obverse die that was fitted into an anvil to create the striations. The meaning of the design is unknown other than to distinguish it from pieces of nonstandard weight. Soon afterward, coinage designs began to take the form of recognizable objects.

Sometime between 700 and 550 B.C., a turtle became the trademark of the coinage (figure 3) from the Saronikon Gulf island of Aegina. Is the reverse punch mark a cross, or perhaps a star? By 480 B.C. the obverse design remained the same, but the reverse punch mark became an incuse square with two crossed lines and a diagonal line (figure 4).

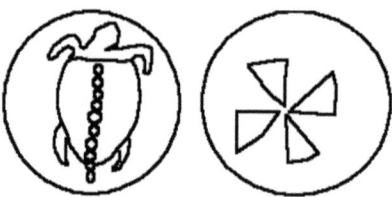   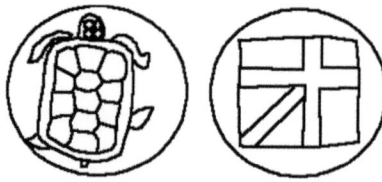

Figure 3. Cross punch mark   **Figure 4. Incuse square**

A triobol, or half drachma, of Aegina was issued during this same era that contained both a crescent and an annulet in the reverse design (figure 5). Certainly the crescent represents an

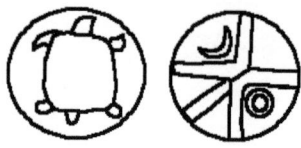

Figure 5. Triobol of Aegina

astronomical body, and therefore the annulet probably does also. These symbols may generically represent the sun (annulet) and the moon (crescent), or perhaps the phases of the moon or the partial and annular phases of a solar eclipse.

Solar Symbols

The most important of the celestial objects is the sun. On ancient and medieval coins the sun or a solar deity was depicted as a star, wheel, annulet, or rayed disk. One of the earliest depictions is found on several ½ staters of Miletos that were struck in the late sixth century B.C. The obverse design is a lion, with a floral star pattern on the reverse (figure 6).

Figure 6. Miletos

Many ancient coins share similar reverse sun-star motifs. Without additional information or more accurate dating of the coin, determining if the symbol represents a solar deity or the physical appearance of the sun is difficult, but a solar reference is most likely the inspiration for the designs (figures 7–10).

 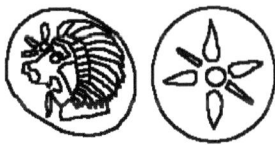

Figure 7. Mysia (200–100 B.C.) **Figure 8. Caria (395–377 B.C.)**

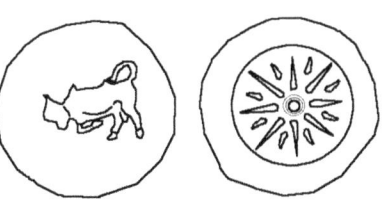 

Figure 9. Kampanoi (c. 344–336 B.C.) **Figure 10. Polemo of Pontus (36 B.C.)**

As the workmanship of the designs increased in complexity, the solar symbol was often reduced to a small star that was included as part of the design or added as an additional mark in the field (all but the outer legend) of the coin (figure 11).

In the Greek city of Uranopolis in Macedonia, the sun was depicted as a large disk with many small rays emanating from it (figure 12). On the other side of the coin, however, the sun is portrayed in its stellar form as a central pellet with large rays. Seated next to the sun is Aphrodite Urania. Some evidence indicates that she may have been a moon goddess. A smaller coin of Uranopolis has a star and crescent that represent the sun and the moon, thereby reinforcing the notion that Aphrodite and the star may represent the sun and the moon.

Figure 11. Miletos (300–250 B.C.) Figure 12. Uranopolis (c. 300 B.C.)

A wheel was often used to represent the sun or the revolution of the heavens. The Celtic cross, a cross within a circle that looks like a wheel, developed in the Carpathian region around 3000 B.C. as a sun symbol and slowly spread across Europe. One of the earliest solar wheel designs appeared on a small silver diobol of Mesembria, Thrace, struck between 450 and 350 B.C. (figure 13). The obverse depicts a crested helmet and the reverse a radiate wheel. The rays surrounding the wheel clearly represent the sun. Many solar wheel designs can be found on ancient coins from different lands. In some cases the date of the coin is known well enough to associate the motif with a total or annular solar eclipse.

In Apulia, some time during the second half of the third century B.C., a coin with a helmeted Athena on the obverse has a prominent star within an annulet on the reverse (figure 14). Here, the rays are much more star-like than the spokes of a wheel, although this design is usually described as a chariot wheel. The pellets above Athena represent the denomination of the coin, a quincunx.

Figure 13. Thrace Figure 14. Apulia

Celtic pieces also offer some evidence that wheel-like structures are representations of the sun. The earliest coins of Britain were made in Gaul and brought to Britain during the migration of Belgic peoples. These issues became the basis for the designs of Celtic coinage beginning around 95 B.C. Many of the Celtic coins contain astronomical symbols. Consider two gold staters struck between 55 and 45 B.C. (figure 15). On the stater of Corieltauvi, a star-shaped figure and a crescent are found below a horse. These figures are presumably the sun and the moon. On a stater of Atrebates that was struck during the same period, a crescent and wheel-like feature are shown under a horse. Although this is not conclusive evidence, the relative positions of the wheel and crescent are identical to those of the Corieltauvi stater. This may indicate that one stater

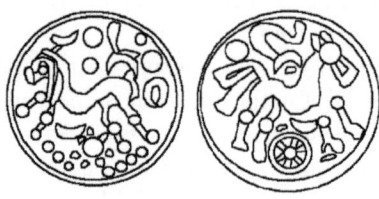

Figure 15. Celtic staters

was used as the model for the other, with the die maker copying the figures in the die, and thus reversing the figures on the coin.

Baldwin (1915) hypothesized that various forms of triskeles, swastikas, and whorls on ancient coins were used to represent solar motion or a wheel. These motifs are found on many ancient coins and other artifacts (figure 16).

Figure 16. Triskeles (Lycia, fifth century B.C.); swastika (Thaena, first century B.C.); whorl (Macedonia, 196–168 B.C.)

During the Roman and Byzantine periods, star symbols continued to be used often as both major and minor design elements. Usually the star represented a solar deity, a heraldic symbol derived from a solar entity, or sometimes a solar event such as an eclipse (figures 17–18). However, the star as a coin symbol was also used to represent comets, planets, and actual stars. The wheel as a solar symbol was gradually eliminated, and annulets re-emerged early in the medieval period (figure 19).

Figure 17. Roman Empire (318–319)   Figure 18. Byzantium (527-565)

Figure 19. Anglo-Saxon sceat with a bird, two annulets (c. 690–710)

By early in the second half of the medieval period, western Europe had fully embraced Christianity. Stars and annulets are often found on medieval coins (figure 20), especially those struck in feudal regions. While pagan deities were obviously not depicted on these coins, generic

Figure 20. Dinero of Alfonso I of Aragon struck in Toledo for Castile in 1109–11; stars are in two quadrants of the reverse cross, and an annulet is found in both legends

astronomical objects and actual celestial events were. Such symbols may have designated mints, moneyers, or minor changes in coin weight or metallic fineness in addition to representing celestial omens.

Annulets on medieval coins, especially those of western Europe, may represent a ring used in the Christian coronation of a king, but more likely represent the sun. In the 11th century in the feudal French region of Champagne, bishops of Langres stuck deniers that feature a bishop's crozier; a crescent; and either a star, mullet, or an annulet with rays (figure 21). The crescent may represent either the moon or a partial solar eclipse, and the stellar patterns are most likely the sun or an actual solar eclipse.

Figure 21. Deniers of Langres

Eastern mints such as that of Armenia in Cilicia, continued to use Byzantine design styles. On trams of Hetoum and Zabel (1226–71), symbols such as pellets, pelleted annulets, and stars, are often found on the base of the staff between Hetoum and Zabel (figure 22). The reverses of these trams portray a lion, along with stars, crescents, and other symbols.

Figure 22. Armentian Trams

The chapters that follow will show that for medieval Christian Europe, when pagan deities were no longer of concern, solar symbols were used primarily to represent actual solar events such as eclipses, or that a particular solar symbol was immobilized from a coin depicting an earlier solar event.

Lunar Symbols

The moon was also a significant heavenly body for the ancients. The phases of the moon provided a mechanism for recording time, and lunar calendars prevailed. The crescent symbol was used to represent the moon (figure 23). Sometimes a crescent symbolized a partial solar eclipse, which is, of course, a much rarer event than the ever-present lunar phases.

The crescent could represent the actual moon or a deity. The moon was also taken as an omen for earthly events. At the Battle of Marathon in 490 B.C., under a waning moon, the greatly outnumbered Athenians drove back the invading Persians to save Athens. On the classic owl tetradrachm of Athens, a crescent moon appears over the shoulder of the owl on coins struck sometime after the battle as a commemoration to the moon (figure 24).

1. The Sun and the Moon 15

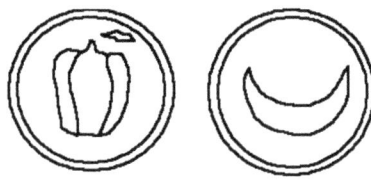

Figure 23. Melos (500–416 B.C.)

Figure 24. Athens (fifth century B.C.)

On some medieval coins, the annulet symbol may represent the moon. Consider a denier of the French province of Languedoc that was struck between 1224 and 1265 by the bishops of the city of Cahor's (figure 25). On one side of the coin, a bishop's crozier appears along with a star and crescent, most likely depictions of the sun and the moon. On the other side of the coin, however, a star and an annulet are shown in the quadrants of the cross. The legends on both sides on the coin contain an annulet. On this coin the annulet may represent the moon, unless the crescent and annulet represent the partial and annular phases of a solar eclipse. It is curious that the moon would be represented by a crescent and an annulet on the same coin, but Languedoc had not been in the path of an annular solar eclipse since 1207, and was not again until 1270, and thus the depiction of an annular eclipse on this coin is unlikely.

A century later, in the French province of Metz, Bishop Adhémar de Montil struck a denier with alternating stars and annulets in the quadrants of the reverse cross (figure 26). On this coin, the eyes and nose of the "man in the moon" face actually appear in a circular representation of the crescent moon.

Figure 25. Denier of Cahors

Figure 26. Denier of Metz

Although the crescent and annulet representations of the moon are the most widely used lunar depictions on ancient and medieval coinage, symbolic representations of lunar deities exist as well, such as Aphrodite Urania representing a moon goddess on Greek coinage (figure 12).

An astronomical engraving (figure 27) from the Seleucid period (301–164 B.C.) in Mesopotamia is interpreted by scholars to include the constellation Taurus (the bull), the moon, and a seven-star cluster. The star cluster is presumably the Pleiades. Note that the moon contains a deity fighting a lion. The moon sometimes passes in front of the Pleiades star cluster

Figure 27. Seleucid engraving

forming a splendid astronomical sight, especially if it is in a thin crescent phase, and this engraving may well represent an occultation of the Pleiades by the lunar disk.

Now consider a similar design on a ⅛ shekel (figure 28) of the Phoenician city of Sidon (372–358 B.C.). Here we see a man or deity battling a lion. Was this a symbol of the moon as well? Modern man looks at the dark lunar maria against the brighter surface of the moon and visualizes a face, or "the man in the moon." By using a little imagination, the deity and lion from the coin of ancient Sidon can be aligned with the lunar maria (figure 29).

Figure 28. Sidon

Figure 29. The moon

A common motif in Greek design is a portrayal of Hercules slaying the Nemean lion (figure 30). After discarding his club, Hercules defeated the lion with his bare hands, and then wore its skin as his garment and used its head as his helmet. The mythical representation of this struggle bears a remarkable resemblance to the Seleucid engraving and the coinage of Sidon. Greek coinage of the period often depicts the head of Hercules with his lion helmet (figure 31), and this motif was used on Seleucid coinage.

Figure 30. Hercules slaying the lion

Figure 31. Seleucid tetradrachm

Generic solar and lunar symbols alone or in combination in coinage design do not offer much opportunity for correlation with actual celestial events; however, these symbols in combination with those that represent stars or planets may depict heavenly groupings actually witnessed by ancient and medieval man that can be re-created on modern computers. Therefore the representation of stars and planets on coinage should be investigated.

The Stars and the Planets

Ancient Rome: Jupiter with seven stars or seven planets

In ancient times man knew of seven planets. These objects were the heavenly bodies that changed their positions with respect to the stars. The word *planet* is derived from the Greek word *planetas*, which means *wanderer*. In addition to the sun and the moon, which were referred to as planets, Mercury, Venus, Mars, Jupiter, and Saturn comprised all the planets visible to the naked eye.

The designs of many ancient coins include one or more stars. While a single star usually represents the sun, a deity, or sometimes a comet, multiple stars may represent either stars or planets. Two stars on ancient coins, however, usually represent the Dioscuri (Castor and Pollux), the two mythical brothers placed by Zeus into the constellation of Gemini. The Romans believed that Castor and Pollux had helped them defeat the Latins in the Battle of Lake Regillus (496 B.C.). Sometimes they are depicted by busts (figure 32); at other times only by their star crowned, egg-shaped helmets; and sometimes just the two stars are portrayed.

Figure 32. The Dioscuri

Without any specific event to explain the design, the most probable explanation for seven stars is a depiction of the seven planets. A less likely, but possible, explanation would be the Pleiades star cluster. The number seven was holy to ancient man, and the seven stars of Ursa Major (the Big Dipper) had a special status in mythology. The Pleiades star cluster, or the Seven Sisters, was also special. The seasonal disappearance of the Pleiades signaled the time to plant and its reappearance signaled the time to harvest. Even though only six stars are easily visible in the Pleiades, various legends from different ancient lands explained the missing star. The ancient Greeks identified the missing sister as Merope, the one who married a mortal and had to hide her face in the heavens.

The attribution of planets and stars on ancient and medieval coins requires quite a bit of analysis. For a few coins, some numismatic scholars have interpreted stars to represent geographical territories and divisions of the church, although these more earthly attributions are somewhat questionable, as such arguments are poorly supported. To make the interpretation more confusing, pellets were sometimes used to represent the planets in addition to their more common use as decorations and as a mark of denomination.

Planets Represented by Stars

Quite often, planets or planetary deities such as Mars were represented by one or more stars in coinage design. Consider for example, a tetradrachm of Uranopolis (figure 33), where the solar disk found on the didrachm design is now surrounded by a crescent moon and five stars. The stars are certainly the five known planets that wander in the heavens along with the sun and the moon.

Coinage of the Gaulish tribes began in the late second century B.C. and their designs were heavily influenced by the style and high quality of earlier Greek coinage. Profile busts and horses

2. The Stars and the Planets 19

Figure 33. Uranopolis

Figure 34. Treveri stater

dominated the Gaulish issues, and a variety of secondary symbols, such as stars, pellets, annulets, and annulets enclosing a pellet accompanied the central designs. Consider a gold stater of the Treveri tribe (figure 34). The obverse of the coin has six disk shaped stars and a wheel enclosed within a large *V*-shaped symbol. The wheel is a sun symbol, and the six stars most likely represent the moon and the five visible planets.

The globe with celestial arcs was a common motif of the ancient religious cult of Mithraism, which began its spread throughout the Roman Empire in the first century A.D. The seven stars that surround an infant atop such a globe on a Roman issue of Domitia (figure 35) are most likely a representation of the seven planets, or possibly the seven stars that comprise Ursa Major. The globe represents the cosmos, on which two arcs cross to form an *X*. One of the arcs is the Zodiac and the other is the celestial equator. The crossing point is an equinox. The infant on the globe may be the dead son of Domitia, represented by Jupiter as a young child. According to mythology, Ursa Major and Ursa Minor (the Little Dipper) were the wet nurses, Helike and Cynosure, who were placed in the heavens as constellations as a reward for raising Jupiter.

Figure 35. Domitia (82–96 A.D.)

Planets Represented by Pellets

Pellets were sometimes used as representations of the planets. In the eastern religions, soldiers often hung circles and disks on military vexilla, or standards, to represent the seven planets. Many examples of ancient coins exist where seven pellets are found in a grouping, and are most likely a depiction of the seven planets (figure 36).

Figure 36. Coritani (first century B.C.)

Pellets were used as planetary symbols on coinage as early as a series of coins of Knossos in Crete. During the period of 500–300 B.C., silver staters were struck at Knossos with reverse designs that depicted the mythical labyrinth Daedalus designed to imprison the legendary Minotaur. Two such coins featured the Minotaur on the obverse (figures 37–38). At the center of the labyrinth on one of the coins is a motif of five pellets. The other has the sun in the center, with five pellets in each of the other labyrinth rooms.

Figure 37. Knossos: five pellets   Figure 38. Knossos: star

Another series of Knossos staters of this period depicted Demeter or Persephone on the obverse and had either the sun or five pellets in the center of the labyrinth, and a third example added a crescent to the outside of the labyrinth (figures 39–41). Clearly the star and crescent symbols depict the sun and the moon, respectively, and thus the five pellets would also symbolize something of equal celestial importance: the five visible planets.

Figure 39. Knossos: sun   Figure 40. Knossos: five pellets

Figure 41. Knossos: star and crescent

A pair of Celtic coins (c.10–61 A.D.) supports the argument that pellets may represent planets. A crudely shaped horse is found on the reverse of each coin. Above the horse on one of the coins (figure 42) six pellets surround a central pellet forming a stellar pattern; however, the seven pellets may not be a crude representation of a single star, but rather, the seven planets. On the other Celtic coin with the same obverse design (figure 43), indicating that the reverse dies are related, the seven pellets are arranged in two rows. This demonstrates that the number of pellets, seven, is the significant concept, rather than the arrangement of the pellets.

During the Roman period, some coins are found with pellets on a standard. The pellets most likely represent the planets or a planetary conjunction. For example, Gallienus (253–268) struck a bronze coin for provincial use in Heliopolis that depicts a bust of the sun god Helios resting

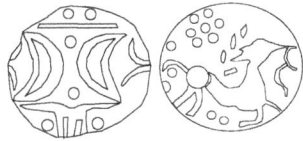
Figure 42. Celtic stater

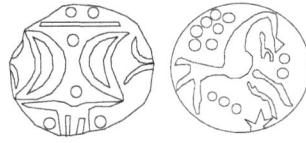
Figure 43. Seven pellets in two rows

Figure 44. Helios with standards

on a throne (figure 44). A standard is located on both sides of the throne, each of which displays a *quincunx*, or five pellets. The top of the standard pole is crowned with a globe, and a globe is found at each end of the cross bar. In the first year of Gallienus's reign, all five of the known planets came together in a loose 20° conjunction in Capricorn and Sagittarius.

To understand the relationship of the planets to the standard, one must consider the history of the *vexillum*, the ancient standard of the Roman army. Religious *vexilla* were a synthesis of a military *vexillum* and *phalerae*, or adornments such as portraits, crowns, and medallions. Various forms of late second and third century Roman standards have religious counterparts that are closely associated with temples and statues of oriental gods of Syria and Mesopotamia. The religious standards seemed to have been based on contemporary military *vexilla*, perhaps to allow Roman soldiers the opportunity to pray to oriental gods in a Roman-like atmosphere.

Religious standards were sometimes found on Roman provincial coinage as well. For example, on some coins struck in Hierapolis during the first half of the third century by Caracalla, Severus Alexander, and Julia Mammaea, a single standard is found between statues of two seated deities, Hadad and Atargatis. In the temple of Atargatis in Dura, which was a smaller version of the temple of Hierapolis, a relief portrays the same standard and seated deities.

Figure 45. Dura altar

The front of a small, gypsum altar (figure 45) found in Dura shows a standard similar to that on the coin of Heliopolis. The top of the standard is crowned by a crescent moon with two circles, or medallions, attached to it. On the banner, six circles are engraved (which scholars maintain represent precious stones or metal discs). To the left of the standard is a stepped altar with an eagle. To the right, a second eagle is found above a bowl (called a crater). Victory and other symbols are scratched into the right side of the altar.

On the left side of the altar, above an arched frame, another standard has seven circles. On the back of the altar, a high pole is crowned with a crescent and orb. Scholars have interpreted the Dura relief to represent a solar trinity. The center standard represents the main deity, with the eagles above the altar and crater completing the triad. The circles on both standards are interpreted to symbolize the planets.

The standards on this third century altar are strikingly similar to the Helios coin of Gallienus. The *vexilla* on the coin are certainly religious in nature, and not military. Each of the standards may represent the other two members of the Heliopolitan triad. Thus if the Dura relief and the Helios coin are related, then the five pellets are probably planets as well.

Pellets were increasingly used to represent planets as coinage design progressed into the medieval period. The motif of seven pellets in a stellar pattern, perhaps representing the planets and their relationship to the heavenly rotation, or wheel, remained as a persistent theme, especially in eastern mints where pagan beliefs remained strong despite the Roman conversion to Christianity. For example, a coin of Rudrasena III of western Satraps (ancient Persia) depicts the sacred mountain at the center of the world in the reverse design (figure 46). Next to the mountain is a cluster of seven pellets, which must be a representation of the seven planets.

Figure 46. Western Satraps (348–378)   Figure 47. Constantine IX

In Byzantium, eastern mints continued to strike coins depicting religious *vexilla*, some of which contained pellets. Consider those of Constantine IX, who ruled the Byzantine empire from 1042–55. One of his gold coins (figure 47) depicts the emperor holding a *vexillum* that has five pellets on it. This design could be immobilized from prior coinage, be a generic representation of the five visible planets, or be representative of an actual planetary conjunction. Twice during his reign all five of the visible planets came together in loose conjunctions. On December 7, 1045, the planets were rising in the east just before sunrise (figure 48). Two years later they were low in the western sky just after sunset (figure 49).

During the 12th century, Louis VI or Louis VII (or both) of France issued a denier with a crude temple or castle containing three pellets inside and a crescent on each side (figure 50). The crescents are certainly astronomical symbols, and may represent one of four partial solar eclipses that would have been observed in France during that period.

The three pellets in a triangular arrangement are consistent with a similar motif used throughout medieval Europe. Philip II, known as Philip Augustus, or Philip the "God-Given," was consecrated as king of France in 1179 and succeeded to the throne in 1180 following the death of Louis VII. On one of his first coins the number of pellets changed from three to five (figure 51).

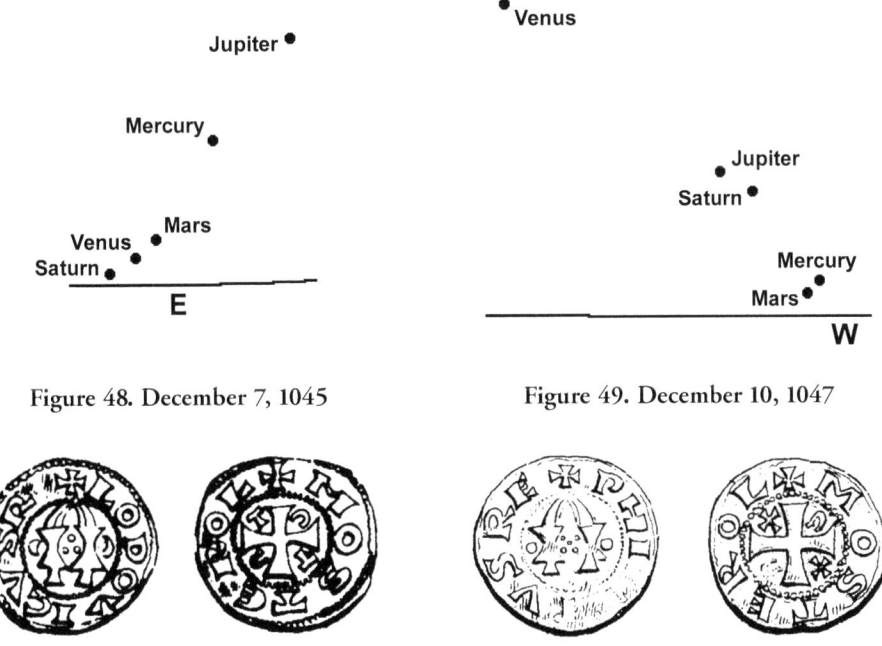

Figure 48. December 7, 1045

Figure 49. December 10, 1047

Figure 50. Crescent and three-pellet denier of Louis VI or Louis VII (1108–80)

Figure 51. Five-pellet denier of Philip II (1180–91)

In 1180 Philip II married Isabelle de Hainaut, the niece of the count of Flanders, Philip d'Alsace, whose wife had died without having children, thereby opening the door for Philip II to claim future rights to northeastern territories. Relations were mostly friendly between the two Philips, and as a result of the Treaty of Boves in 1185, Philip II acquired the county of Amiens. Upon the death of Philip d'Alsace in 1191, the task of unifying northeastern lands under the domain of royal France became much simpler, and through a variety of agreements and confiscations, Philip II greatly expanded the royal domain.

But why was he known as Philip the God-Given, or Philip Augustus, and what does this have to do with the coinage design? In 1184 Arab astrologers in Spain proposed various prophecies in anticipation of the 1186 conjunction of all five of the known planets in Libra. Predictions included great natural and political disasters and benefits for the French.

William, a clerk to the constable of Chester in England, wrote of predictions that a great Christian prince would rise and that England would suffer. The chronicler Roger of Hoveden wrote: "In that year, 1184, the astrologers, both Spanish and Sicilian — and indeed almost all the world's prognosticators, Greek and Latin — wrote much the same prediction about the conjunction of the planets."

All five of the known planets came together within a 5° circle at their closest on September 20, 1186 (figure 52). On that date, they may have been too close to the sun to be easily visible. A week earlier, the five planets would have been seen low in the western sky just after sunset within a 6° separation.

There can be little doubt that the five pellets within the castle on the coin of Philip II can be anything other than representative of the great conjunction of 1186. The astrological significance of the event along with predictions of the rise of a great Christian prince and benefits to the French, would certainly help Philip II consolidate his powers, and perhaps assume the divine title of God-Given.

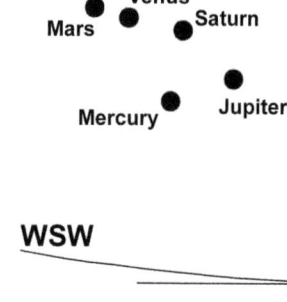

Figure 52. September 20, 1186

Currently no explanation of the title of Philip Augustus has been put forward, other than Philip's assuming the title of a Roman emperor because of his long and powerful reign. Perhaps it refers to a planetary conjunction that foretold the God-given success of an earlier Roman emperor, one who was also a great Christian prince.

The Cross and Three Pellets

The most common pellet design on medieval coins is a triangular arrangement of three pellets. During the early Middle Ages, this motif can be found anywhere on the coin, but in the later medieval period the three pellets are usually within the quadrants formed by a cross. This design transcends across time and political boundaries, and may therefore be associated with a single event common to many medieval cultures (figures 53–58).

Figure 53. Frankish Kingdom (c.650)

Figure 54. Anglo-Saxon England (c.690)

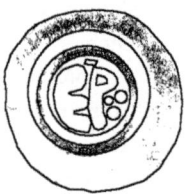

Figure 55. Tuscany (672–688)

Figure 56. France: Charlemagne (768–771)

Immobilization of this three-pellet design over many centuries as a celestial event may be difficult to demonstrate, as such conjunctions occur frequently, although extremely tight groupings are infrequent, though usually visually spectacular. However, if a celestial event were visible at the same time as a significant cultural event occurred, the corresponding celestial design may have become a symbol associated with an immobilized cultural motif.

The persistent immobilization of a design across political cultures would suggest a spiritual

2. The Stars and the Planets   25

Figure 57. Bohemia (967–999)

Figure 58. Ireland (1279–84)

or other nonpolitical reference. As the trio of pellets is often within the quadrants of a cross, or infrequently is part of a bishopric design, significance to the Christian religion is certainly a possibility. Current numismatic thought is that the design represents the Christian trinity, and may or may not be associated with a celestial event. Another possibility is that the motif was adopted as a Christian symbol from an earlier pagan device.

Coinage during the first 500 years of Christianity was that of the Roman Empire, and European coinage of the sixth century saw a gradual transition from designs that imitated Roman and Byzantine coinage to those of the Visigoths, Franks, Lombards, and others who began minting in the last decades of the century. The three-pellet motif is found on many of these coins, suggesting a Roman origin.

A variety of examples of this motif is found on coins of Anglo-Saxon and Norman England, although the persistent immobilization of the cross and three-pellet design on English coinage began with the long cross penny of Henry III in 1247, the period when the church had its greatest influence. Examples of this motif on English coinage are found as late as 1604 on halfpennies of James I.

Consider the petit gros of Bishop Nicolas de Fontaines of Cambrai (1248–72). Not only are the three pellets located within the corners of the cross, but they also appear on both sides of the bishop's headpiece (figure 59).

Variations of pennies (figure 60) struck by the archbishop of Canterbury (833–870) depict either one or three pellets, suggesting that a single pellet and the three pellets may have been used interchangeably, or that the three pellets may represent three entities that may be associated as one.

Figure 59. Cambrai   Figure 60. Canterbury

The triangular arrangement of three pellets or annulets was also a common medieval design element associated with rulers. Several examples exist in contemporary drawings where kings are adorned with the triangular design. For example, a 12th century manuscript illustration shows Holy Roman Emperor Frederick Barbarossa (1152–90) with his sons Henry and Frederick (figure 61). All three have the triangular three annulet design on their robes. Frederick II, son of Henry, was emperor from 1215–50, and held an almost obsessive belief in astrology.

26   Astronomical Symbols on Ancient and Medieval Coins

Figure 61. Emperor Frederick Barbarossa

Evidence of the triangular arrangement of three pellets being a celestial or heavenly event may be found on the 13th century English Psalter map (figure 62). A circular world with Jerusalem at the center is surmounted by Christ and two angels. In the heavens, the stars are clustered into triangular groupings of three pellets.

Figure 62. Psalter map (top portion detail)

A manuscript on St. Paul's Epistles, c. 1164, shows the sun, the moon, and stars along with triangular clusters of three pellets (figure 63). Once again, planets are the logical choice for the pellets, if they indeed represent celestial objects.

Figure 63. St. Paul

Now consider a 10th century Byzantine reliquary found in the Iberian monastery on Mount Athos (figure 64). The piece is covered with a mosaic of precious stones, enamel, gold partitions, and silver filigree. Note the three stones within the corners and at the ends of the cross. This presents an unusual contradiction. If the reliquary is indeed Byzantine, then the three pellets may not necessarily represent the trinity, as the Eastern Orthodox church rejected this concept. If the design represents an actual celestial event, Byzantine designs could still use the motif, but on a historical, rather than a theological basis.

Figure 64. Byzantine reliquary

Note that the triangular arrangement of three pellets is rarely found on Byzantine coinage (figure 65), and the absence of the design is extraordinary compared with the frequency of the design on western medieval coinage.

Figure 65. M. Comnenus of Trebizond (1238–63)

Byzantine coinage is more likely to depict four-pellet, or especially five-pellet, motifs that may represent generic planets or actual planetary conjunctions. Consider the Syracuse solidus of Leontius (c. 695–698). On this coin (figure 66), four pellets in a diamond cluster are found to the left of the reverse cross. The four pellets may represent a four-planet conjunction. In January 694, four of the five known planets came together within a 20° separation, and on December 4, 690, Mercury, Venus, Jupiter, and Saturn came together within a 5° distance.

Figure 66. Solidus of Leonitus

The concept of the Christian trinity was postulated at the Council of Nicaea in 325 and the Council of Constantinople in 381. The adoption of the three-pellet motif as a Christian symbol may well have originated during the reign of Constantine the Great in fourth century Rome (discussed in a later chapter), but examples of its use on coinage struck before the Council of Nicaea rule out the design as a strictly theological device, but rather as an adoption of a pagan symbol that may have represented an earlier celestial event. For example, a Sassanian drachm of Varhran II (276–293) has three pellets in a triangular arrangement added to the fire altar (figure 67). During the first year of his reign, three planets came together in triangular conjunctions on two occasions. On November 8, 276, Mercury, Venus, and Saturn formed a flat triangle in a 3° conjunction. On December 26th of the same year, Mercury, Venus, and Jupiter formed a nearly perfect equilateral triangle in a compact 1° conjunction. This latter conjunction was probably too close to the sun to be visible; however, the ancient Arab astronomers based their astrological predictions not only on visual events, but also on computed events, and their computational abilities may well have allowed them to predict this latter conjunction.

Figure 67. Drachm of Varhran II

A variety of Celtic coins issued between the middle of the first century B.C. and the middle of the first century A.D. contain the three-pellet motif (figures 68–70). A stater of Anted (1–25 A.D.) has three prominent crescents indicating an astronomical reference, along with three pellets in a triangular arrangement and four pellets in a diamond cluster. In addition, triangular arrangements of three annulets are found on both sides of the coin. Three pellets in a triangle are located below a horse's head on staters of Dumno Tigir Seno (10–61 A.D.). A silver unit of Queen Boudicca (61 A.D.) has the triangular arrangement of three pellets below her bust.

Figure 68. Stater of Anted   Figure 69. Stater of Dumno Tigir Seno

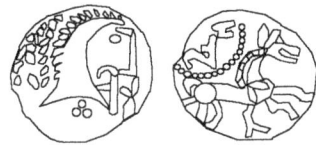

Figure 70. Unit of Queen Boudicca

The three-pellet design must surely have been significant. Although this motif may have been immobilized from earlier staters of the Iceni, a contemporary conjunction may have reinforced the use of the design. As shown on the pennies of the archbishop of Canterbury (figure 60), some of the cross and pellet designs of medieval Europe replace the three pellets with a single pellet. Is it possible that this represents a three-planet conjunction that was extremely tight, or one that may have appeared to some as a single star, and at the same time has religious significance?

Note that some theorize that the Star of Bethlehem was a very close conjunction of two planets that looked almost as if they were one bright star. These theories include the Jupiter-Venus conjunction of June 17, 2 B.C. as a strong contender based on other historical considerations. This conjunction occurred near the bright star Regulus, which could possibly be the third object. Other theories propose that a three-planet conjunction in 6 B.C. is the source.

While my purpose is not to try to identify the Star of Bethlehem, some correlation of the cross and three-pellet design to an astronomical event of importance to Christianity is indeed possible. Ten three-planet conjunctions that were separated by 3° or less occurred between 10 B.C. and 10 A.D. One of the tightest conjunctions (Mercury, Mars, and Jupiter) occurred on August 26, 2 B.C. During this period the most spectacular conjunction of planets occurred when Mercury, Venus, and Mars, rose in an almost perfect equilateral triangle in the astrologically significant eastern sky on November 2, 1 A.D. All three planets fit within a quarter of a thumbnail of an outstretched arm. Nearby Jupiter and a thin, crescent moon formed a remarkable sight (figure 71). Three days later all four planets were in a tight cluster. Could this conjunction be the basis for the crescent and the three-pellet and four-pellet motifs on the Celtic stater of Anted?

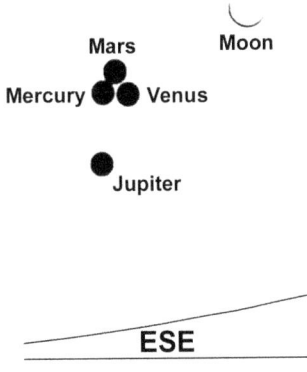

Figure 71. November 2, 1 A.D.

Another example may correlate with these three-planet conjunctions. The Romans made Herod I king of Judea by proclamation in 40 B.C. and he ruled until his death. It was on Herod's order that the three wise men were sent to find the future king of the Jews. The date of Herod's death is questionable, but various theories place it between 4 B.C. and 1 B.C. At some point during his reign, small bronze coins were issued with five pellets (figure 72), suggesting the five planets, and one variety is found with a triangular arrangement of three pellets in the design (figure 73). Could this be a representation of the same planetary symbol found in Christian designs?

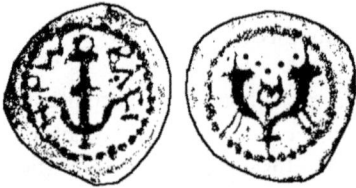

Figure 72. Bronze issue of Herod   Figure 73. Three pellets

The triangular three-pellet motif is, however, found much earlier than at the beginning of Christianity. Three pellets are sometimes shown on ancient Greek coinage, but usually as a mark of denomination, and not as a symbol with other meaning. However, a Roman Republican denarius of L. Calpurnius Piso Frugi, struck in 90 B.C. has the triangle of three pellets behind the bust of Apollo (figure 74). The symbol on this coin can not be a mark of denomination.

Figure 74. Roman Republic 90 B.C.

This motif actually precedes the first known coinage. A carved lapis seal made in Mesopotamia around 1600 B.C. features a griffin surmounted by a triangle of three pellets, with two single pellets behind it (figure 75). The five pellets may be the visible planets. A similar motif (partially damaged) is found on a Minoan seal of 1450–1400 B.C. where three circles are shown on the wing of a griffin and another circle is above the griffin (figure 76). Recall the use of five pellets on the early coinage of Knossos. On an ancient stone of unknown origin, a crescent is found above a griffin, suggesting that the position above the griffin may have been reserved for celestial objects or deities (figure 77).

Figure 75. Mesopotamian seal   Figure 76. Minoan seal

Figure 77. Griffin and crescent

Transition or Conflict?

Thus both stars and pellets were used from the earliest coinage through the medieval period to represent planets, and the triangular configuration of three pellets survived for more than 3000 years, evolving from a pagan to a Christian symbol. Interestingly, this same triangle of three pellets was also used as the ancient symbol for silver, even though it appears on coinage of all metals. A gold coin of the Celtic king Cunobelin (c. 10 A.D.) depicts a Celtic wheel below two horses (figure 78). A single pellet is contained in each quadrant of the wheel. Does this imply that the Christian cross and pellet motif was a transition from the Celtic symbol for the sun or the rotation of the heavens?

During the later middle ages, the three-pellet motif was widely used in the quadrants of a cross, but was sometimes replaced by astronomical symbols. Does this imply a celestial origin for the three pellets? Or does it imply that the quadrants of the cross were reserved for heavenly and celestial motifs? The forces of Christianity and astrology were not in harmony, but in conflict. Perhaps sovereigns were reaching out to followers of both beliefs for their support.

Consider two silver ambrosinos of Milan, both struck between 1250 and 1310. Both coins feature St. Ambrosius enthroned with a crozier. One coin has three pellets on each side of the throne and a trefoil of pellets in the quadrants of the reverse cross (figure 79). The other coin omits the three pellets beside the throne and has crescents in the reverse quadrant in lieu of trefoils (figure 80). The cross in the reverse legend is also absent.

Figure 78. Cunobelin stater

Figure 79. Milan: pellets (1250–1310)

Figure 80. Milan: crescents (1250–1310)

Eclipses

Croatia c.1239

At the same time, A.D. 1239 on the third day from the beginning of the month of June, a wonderful and terrible eclipse of the Sun occurred, for the entire Sun was obscured, and the whole of the clear sky was in darkness. Also stars appeared in the sky as if during the night, and a certain greater star shone beside the Sun on the western side. And such great fear overtook everyone, that just like madmen they ran about to and fro shrieking, thinking that the end of the world had come (reported from Croatia and recorded in Thomae Historia Pontificum Salonitanorum et Spalatinorum *[quoted in Stephenson 1997]).*

Path of the total solar eclipse of June 3, 1239

The geometric interaction of the earth, the sun, and the moon can result in eclipses. When the earth's shadow covers the moon, a lunar eclipse occurs. Although lunar eclipses were often recorded in medieval writings, this type of eclipse is not a rare event for any particular geographic location, and because people can witness such events every few years, the likelihood of a lunar eclipse being depicted on coinage is remote. However, a solar eclipse is not only much rarer, but of all the celestial events visible to ancient man, a total eclipse of the sun would have been the most awesome.

Ancient battles were stopped by such eclipses. Consider the eclipse reported by Thales of Miletus. Thales, an Ionian Greek mathematician and astronomer, was credited 100 years later by the ancient Greek writer Herodotus to have predicted the eclipse of 585 B.C. that stopped the war between the Lydians and the Medes. During a battle in northern Turkey, the two sides immediately stopped fighting when the eclipse occurred, and took the eclipse as a sign that they should make peace.

Consider what the protagonists may have seen in the heavens. For about an hour prior to the main event, the moon slowly took a larger and larger bite out of the sun, until just minutes before totality, when a sunset was seen around the horizon as day became twilight. Then moments before totality, the shadow of the moon raced across the earth and darkened the sky as if a divine curtain had been pulled across the heavens. As totality began, the last rays of sunlight were seen through low points in lunar mountains, and the blackened sun had brilliant diamonds of light on its edge until the entire solar disk was covered, and finally the magnificent solar corona became visible (figure 81).

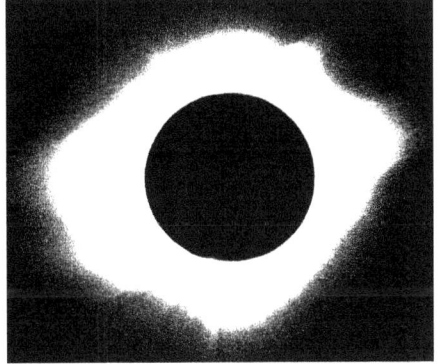

During this process the temperature dropped by as much as 15°F. Wind currents changed. Animals became still and night creatures awoke. The planets and bright stars became visible. A few minutes later totality was over and the solar disk slowly reappeared. Some of those who had been mesmerized by the eclipse and had stared at the thin partial eclipse

Figure 81. Totality

just prior to and after totality, when the brightness of the sun was diminished enough to be comfortable to view, but still dangerous to the eye, found themselves with permanent blind spots in their eyes. No wonder that ancient man took such rare sights as divine omens.

When the distance between the earth and the moon is too great for the lunar disk to completely cover the sun, an annular ring of fire is visible around the sun (figure 82). While not as spectacular as a total eclipse, almost all of the phenomena leading up to totality are observable.

On average, about 23 solar eclipses are visible somewhere on the Earth per decade, 8 of which are annular and 7 are total, with the rest seen only as partial. However, a total solar eclipse is only visible over a fraction of 1 percent of the earth, and for any given location, a total solar eclipse occurs only once every four centuries, therefore the chance of witnessing such an event was extremely rare. Annular eclipses are

Figure 82. Annular Eclipse

somewhat more frequent, but still quite rare. More frequently, outside the path of totality a partial solar eclipse is visible where part of the solar disk is blocked by the moon and the sun appears as a crescent. However, because the sun is so bright, the sky will appear to darken for only a few hundred miles away from the region of totality, where the sun would appear as an extremely thin crescent. While the geometric mechanisms of eclipses were well understood in medieval times, the general inability to predict such occurrences would have prohibited noticing all but the thinnest partial eclipses because of their brightness, or if they were observed close to the horizon, where the rising or setting sun is greatly dimmed by the increased light path through the atmosphere and solar crescents could be comfortably observed.

Some may question how often a partial solar eclipse away from the horizon would be visible without modern viewing accessories, such as filters, cameras, and telescopes. As the solar disk is covered by the moon, the sky darkens, the atmosphere may become still, the temperature drops, and animals may respond with unusual behavior. The combination of these effects draws attention to the fact that a change is in process. At this point, nature can provide the equivalent of a modern camera.

When the atmosphere is still, small holes in leaves on trees perform like hundreds of pinhole cameras, with images of the partially eclipsed sun projected on any flat surface (figure 83). Thus when the sun was low in the sky and seen though the dense atmosphere, or when leaves were on the trees to act as pinhole cameras, partial solar eclipses would have had a better chance of being observed than midday winter eclipses.

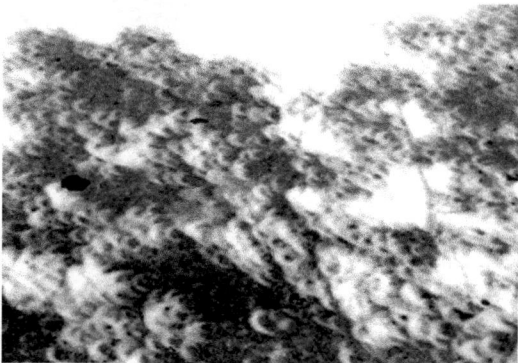

Figure 83. Tree shadow showing hundreds of small crescents during the partial phase of the annular solar eclipse of May 10, 1994

Two other factors in the observation of solar eclipses must be considered. The first concerns the rotation rate of the Earth, or in more simple terms, the length of a day. Tidal frictions cause the rotation of the Earth to slow by a slight amount every day, and each day becomes a bit longer than the previous one. While this is generally unnoticeable, the cumulative effect over hundreds of years affects what portion of the Earth is in the lunar shadow during a solar eclipse.

The change in the rotation rate of the Earth for the past three hundred years or so is fairly well known, but for dates prior to the mid–1600s, calculations pertaining to the change in rotation are based on theoretical models and dates and locations of recorded ancient and medieval eclipses (see Appendix A for more technical details). Uncertainty in the rotation rate value for the second half of the medieval period results in inaccuracies in computer re-created eclipse paths on the order of 50 to 200 miles, which is not too much of a problem; however, the inaccuracies in re-created ancient eclipse paths can be considerable.

The second factor is the weather. Unlike planetary conjunctions that evolve over a number of days, or comets that may be visible for weeks, solar eclipses last for only a few hours, with the length of totality measured in minutes. Rain and clouds can easily obstruct the viewing of a solar eclipse. Therefore, having historical written records of observed eclipses is desirable.

On ancient and medieval coins total solar eclipses were represented by star symbols that stood for the sun, or by mullets, which are stars with a central hole (or less frequently a central pellet). Some coins actually show the solar disk with rays and a central hole. The mullet design is an excellent representation of the solar corona and the lunar obstruction as seen during a total eclipse. Annular eclipses were represented by annulets and generic star symbols, and for both total and annular solar eclipses, the partial phases before and after totality or annularity were represented by crescents. Thus some confusion can arise between the representation of eclipses and of other celestial objects for which the same symbols were used.

Ancient Solar Eclipses

While eclipses were featured on ancient coins, the mullet device rarely appeared until the second half of the Middle Ages. Thus the sun symbol, used to represent an eclipse in ancient coinage design, might refer to the physical appearance of the sun, or perhaps to the sun god, who caused the sun to darken and then to reappear. One of the earliest uses of a mullet is found on a Celtic Trinovantian quarter stater (figure 84) struck around 55–20 B.C. Solar eclipses crossed the British Isles in 40 B.C. and 35 B.C. and may be represented on this coin.

Figure 84. Trinovantian quarter stater

Numerous examples exist where a sun symbol may be associated with an ancient solar eclipse. Between 76 and 78 A.D., the Roman Emperor Vespasian issued a silver denarius with a large star above the prow of a ship (figure 85). The prow design usually represented a victory at sea. On January 5, 75, a total solar eclipse crossed the Mediterranean Sea and Italy just south of Rome. A strong case can be made that this eclipse is represented on the coin, although no records exist of a significant naval victory in that year.

The star above the prow of the ship as an omen for victory did not originate with Vespasian.

Figure 85. Vespasian

Examples are found on Phoenician shekels as early as 345 B.C. In the Macedonian city of Amphipolis, a tetrobol was issued sometime between 196 and 168 B.C. with a star above the prow of a ship (figure 86). On the obverse of this coin, four crescents and a pellet form a star, or a sun symbol, in the center of a shield. In 192 B.C. Syrian forces under the command of Antiochus III invaded Greece. A year later Roman forces routed Antiochus III at Thermopylae, and the Syrian forces pulled back to Asia. In the winter of 190 B.C. Roman forces once again defeated Antiochus III at the Battle of Magnesia. Prior to this battle, the Roman navy, aided by the fleet of Rhodes, destroyed the Syrian fleet off the coast of Crete. Surely this Macedonian tetrobol commemorates an important naval victory. A total solar eclipse crossed Macedonia in March 190 B.C., and the star on the coin may represent this eclipse.

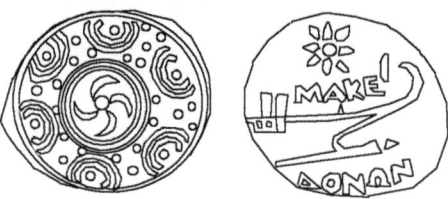

Figure 86. Tetrobol of Amphipolis

In Macedonia, around 196–179 B.C., a small, silver coin was issued with a Macedonian shield on the obverse and a cavalry helmet on the reverse (figure 87). Beneath the helmet is a star. Does this coin depict the Battle of Magnesia and the 190 B.C. solar eclipse?

Figure 87. Macedonian helmet

Riders on horseback were often used to represent warriors. On a coin of Calabria (Italy) struck between 281 and 272 B.C. the obverse has a rider on horseback (figure 88). The reverse of the coin has Taras astride a dolphin, with stars on both sides. Taras was the founder of Tarentum in Italy, and legend states that a dolphin rescued him from a shipwreck. On August 6, 282 B.C., a total solar eclipse crossed Italy. Perhaps the stars represent this eclipse.

Figure 88. Calabria

Second century B.C. coins from eastern Gaul show an annulet in front of a rider on horseback or a star below the rider (figures 89–90). Although the dates of the coins are have not been firmly established, total solar eclipses crossed the area in 190, 188, 174, and 163 B.C., and an annular eclipse crossed in 105 B.C.

Figure 89. Eastern Gaul (annulet)

Figure 90. Eastern Gaul (star)

One of the coins of Alexander Jannaeus, king and high prince of Judea from 104 or 103 to 76 B.C., depicts a star surrounded by an annulet (figure 91). An annular solar eclipse crossed Judea in September 104 B.C. which was likely taken as an omen of the death of Judah Aristobulus I and the succession of Alexander Jannaeus.

Figure 91. Alexander Jannaeus

The most obvious interpretation of a crescent symbol in coinage design is that it represents the moon. However, other explanations for the addition of a crescent to coinage design may have been more significant. A thin crescent sun formed during a partial solar eclipse may have been viewed as a divine omen. One approach to demonstrate the significance of a crescent symbol is to ascertain whether a significant historical event, a partial solar eclipse, and a coinage design change occurred at the same time.

Such an example is found on the ancient coinage of Boeotia. The Boeotian League was a loose federation of Greek cities dominated by Thebes and situated just to the north of its arch rival, Athens. Boeotia was often its own worst enemy, with civil wars that kept its population in a constant state of turmoil and antipathy toward it by its neighbors.

At the conclusion of military engagements against the Spartans in 387 B.C. the Boeotian confederacy was dissolved by its members with each city then having its autonomy under the leadership of Thebes. Separatists desired Spartan alliances, and in 382 B.C., the Theban citadel was delivered into Spartan hands and the opposition was exiled. In 379 B.C. Thebes was liberated, the Spartans were expelled, and the Boeotian League was re-established. For the next eight years other Boeotian cities were taken from Spartan hands and once again became part of the league.

Beginning with its first coinage struck around 600 B.C., a simple Boeotian shield was the primary obverse design for 400 years. After the liberation of Thebes in 379 B.C., a new federal currency was issued. Some of the Boeotian coins issued after the liberation contained a new feature: a crescent (figure 92).

On November 5, 380 B.C. a solar eclipse crossed the Mediterranean Sea and north Africa and

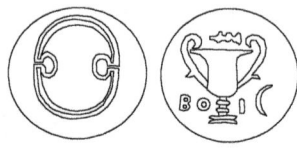

Figure 92. Amphora and crescent

would have been seen as an extremely thin crescent in Greece. Another solar eclipse crossed Phoenicia on May 2, 379 B.C., and also would have appeared to the Boeotians as a crescent. Perhaps either or both these events were taken as an omen for the liberation of Thebes. Under the assumption that Aphrodite was a moon goddess, her appearance with two crescents (figure 93) might signify the interaction of the sun and the moon during both events. According to Plutarch, the mechanism of solar eclipses was well understood by the time of the Peloponnesian War (431–404 B.C.).

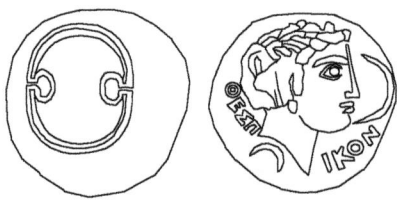

Figure 93. Aphrodite and two crescents

One possibility is that the Boeotians copied the crescent design from the waning moon and owl coinage of their Athenian neighbor, but Attica had used crescents on its coinage for more than a century without being copied by the Boeotians. Furthermore, Athens was usually a combatant neighbor rather than a friendly one. A more likely explanation is that the Boeotians chose a new celestial event to commemorate their victory at Thebes, and on one of the new issues, the reverse design contained another new feature: the sun, perhaps signifying their representation of a solar event.

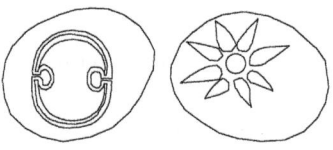

Figure 94. Shield and sun

Medieval Solar Eclipses

As Christianity spread throughout medieval Europe, ancient pagan beliefs in celestial gods that were represented by the planets gradually evolved into astrological predictions based on planetary observations and other celestial events that were in conflict with the teachings of the church. The general populace and sovereigns regarded comets and solar eclipses as omens for battle victories and changing kingdoms. Depictions of astrologically significant planetary conjunctions in the form of pellets and stars were sometimes used on medieval coinage, but stylized, and often crude, representations of actual comets and solar eclipses emerged as the predominant celestial symbols.

The number of medieval eclipses and examples of coinage with solar eclipse symbols are too numerous to present all of them here. Instead, this section discusses a selection of medieval solar eclipse events for which new coins issued along the eclipse path have astronomical symbols that most likely represent the celestial event. The historical context for some of these coins will be discussed in detail, either in this chapter or in subsequent chapters. Appendix B provides additional examples.

Each eclipse path figure shown here includes a depiction of the eclipse, the central eclipse path and boundaries where totality (or annularity) could have been observed based on a value for the variation of the earth's rotation rate adopted by the author (see appendix A), the central eclipse path (dashed line) using a standard model for the Earth's rotation rate, and lines where the eclipse was seen as a 75 percent partial eclipse. Each figure shows places of origin for coins that are discussed. In addition, some of the figures include small insets that show the appearance of the eclipse in London, Bordeaux, and Rome for comparison with regions in or near the eclipse path. Mullets, stars, annulets, and crescents that symbolized the eclipse as seen from various sites were added to local coinage to reinforce their divine messages.

Numismatic evidence supports the notion of some harmony between solar and cometary omens and church teachings, or perhaps beliefs in omens were so strong that the church had to accept them. Consider, for example, the early 14th century issues of Cologne. Archbishop Walram of Julich (1332–49) issued a penny with mullets (figure 95) and a groschen (figure 96) that has a bust with three annulets in the border pattern and on the bishop's hat, and mullets at the beginning and end of the obverse legend. Varieties of this type are found with mullets on the bishop's hat. The mullet (or star) and cross reverse on the Cologne penny was an immobilized design that had been in use for more than century and may have been reinforced by various eclipses that crossed northern Germany. In 1330 a total solar eclipse passed just to the north of Cologne (figure 97), and could certainly have been be the basis for the mullets in the coinage of Walram of Julich.

Figure 95. Penny of Walram of Julich

The use of mullets to represent total solar eclipses continued into the late 17th century. An excellent example is found on coinage of the French province of Nevers. Consider the issues of Charles II de Gonzague, count of Rethel (1621–37). Rethel passed to the House of Gonzague in 1564 when Henriette de Cleves, duchess of Nevers and countess of Rethel, married Louis de Gonzague, son of Frederic II, duke of Mantoue. Their son

Figure 96. Groschen of Cologne

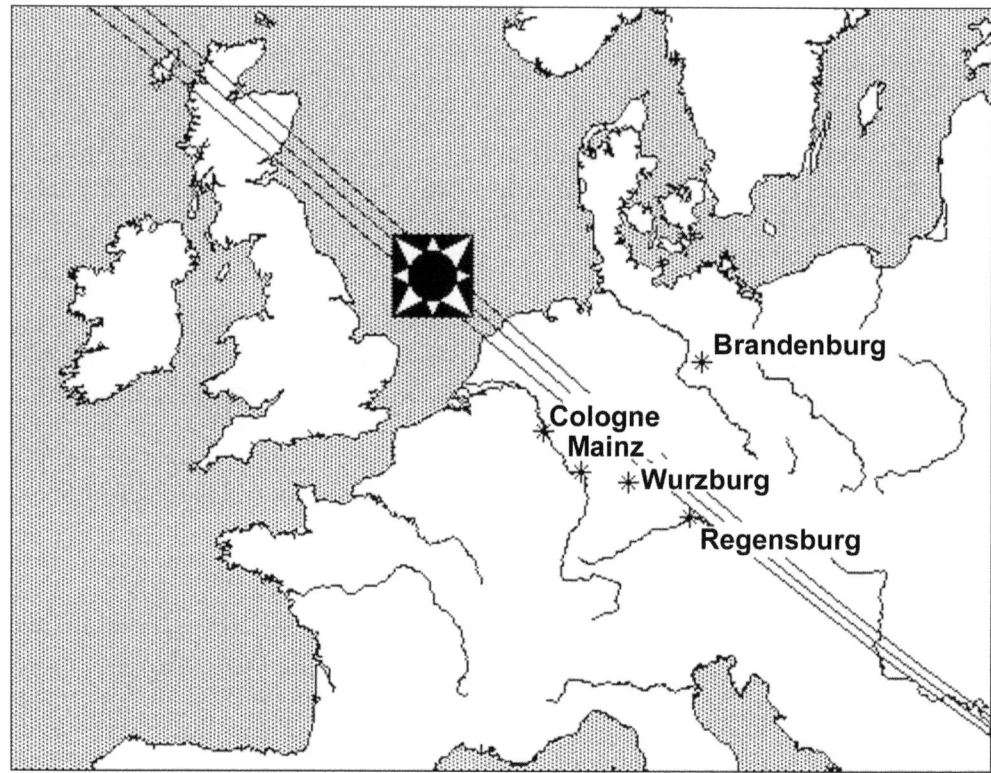

Figure 97. Path of the July 16, 1330, total solar eclipse

Charles II, duke of Nevers and Rethel, and prince of Arches, issued coinage at three mints, Rethel; Charleville, which he founded in 1609; and Arches. Mullets are found on the coinage of Nevers that matches the date of an eclipse. On June 10, 1630, a total solar eclipse crossed France just to the south of Nevers, and would have been seen as a thin crescent farther north in Rethel (figure 98).

As early as 1631, double liards were issued with a mullet in the middle of three fleur de lis (figure 99). The same type was also issued without the mullet. Perhaps the issue with the mullet was for use in Nevers, and those without were minted in Rethel and Charleville. There can be little doubt that these mullet issues were representative of the eclipse.

Earlier issues of Charles II portray his holding either a bow (arch) or the sun in his hand (figure 100), and his thalers portray an eagle with a sun on the shield. These representations of the sun are not as a mullet. The sun symbol had played a significant role in the coinage of Nevers as an immobilized design since 1199, and was perhaps a reference to the total solar eclipse of 1178.

Some may argue that the mullet is merely a representation of the sun and not an eclipse; however, in nearby Orleans, Gaston, the second son of Henri IV, issued a jeton (token) in 1631 that is clearly a representation of the total solar eclipse (figure 101). A full moon covers the solar disk, and solar rays extend from behind the lunar face. The geographical proximity of the two regions, and the issuance of these two coins in the same year, adds additional evidence to the Nevers mullet being representative of the eclipse.

In his 1936 catalog of jetons, Pradel provided a symbolic interpretation of the moon and the

3. Eclipses 41

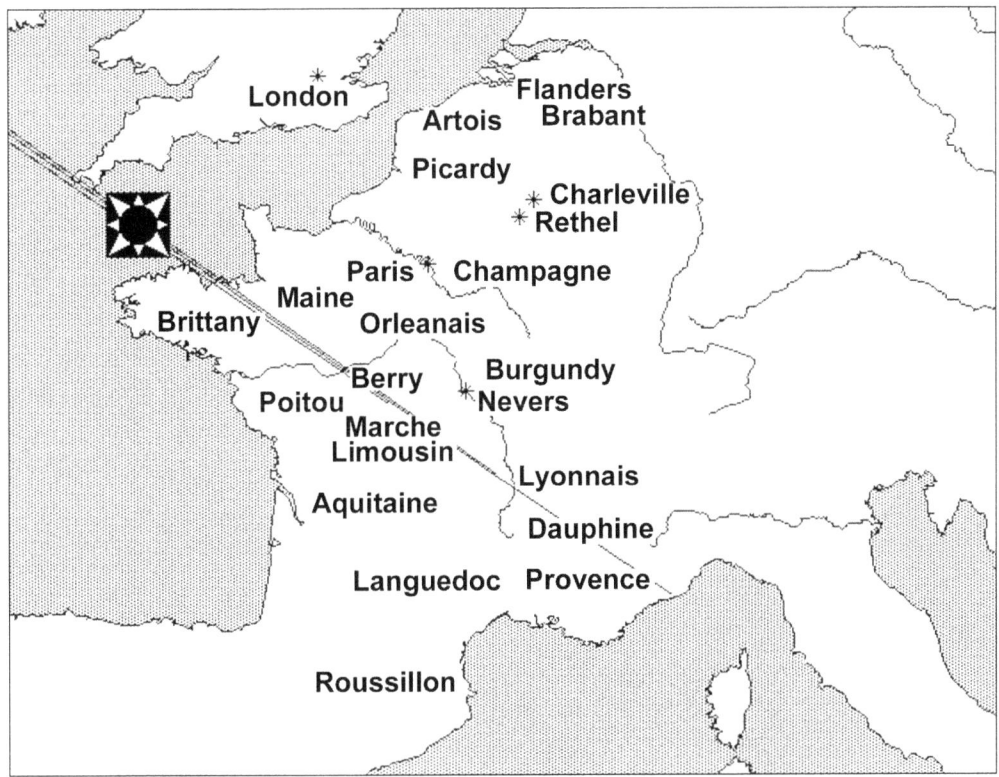

Figure 98. Path of the June 10, 1630, total solar eclipse

Figure 99. Post–1630 double liards of Charles II

Figure 100. Early issues of Charles II

Figure 101. 1631 jeton

sun on this jeton. According to Pradel, Gaston's jeton symbolized the honor and glory of being the second most important person in France, while his brother the crown prince, represented by the sun, was the most important. The total solar eclipse seems a more likely explanation.

The Annular Eclipse of 1084

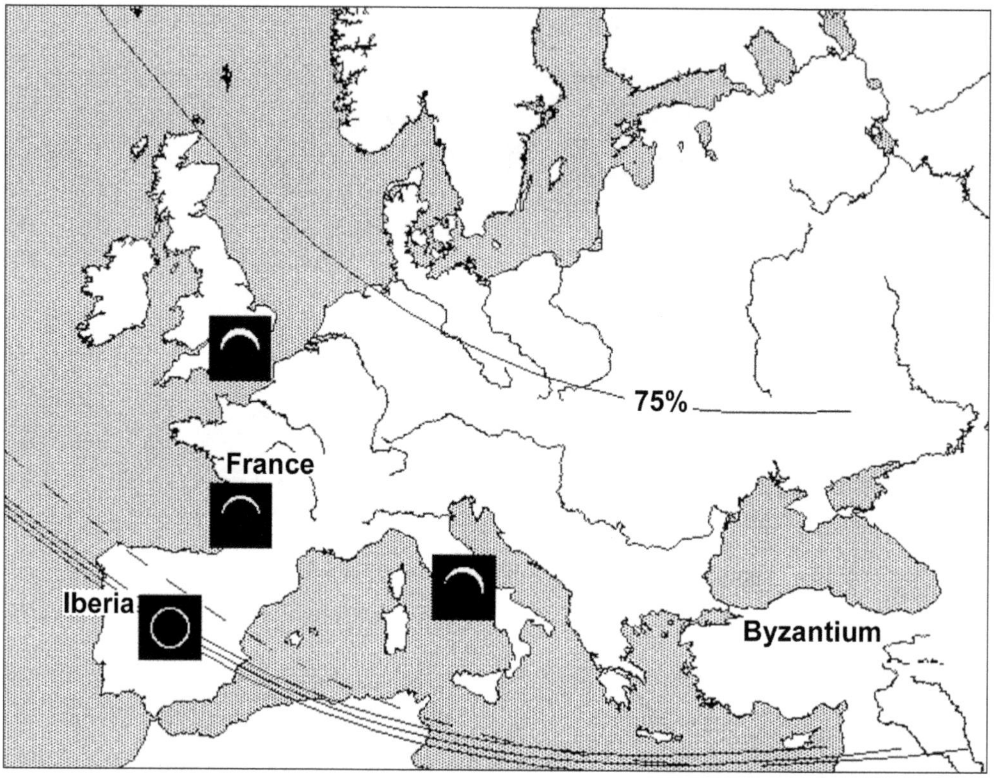

Figure 102. Path of the October 2, 1084, annular solar eclipse

The annular eclipse of 1084 was fairly narrow, with a path of annularity that ranged from 63 to 122 kilometers in width (figure 102). The eclipse took place between 12:37 and 14:52 UT (Universal Time, the time in Greenwich, England). At its maximum view, annularity lasted for only 1 minute and 24 seconds.

The Iberian Peninsula

Early in the eighth century, Arabs from North Africa invaded the Iberian peninsula. For the next 700 years, the *Reconquista* (reconquest) of the land from the Muslims (Moors) became the predominant theme of medieval Spanish and Portuguese history. In the north, Christian resistance developed into the kingdom of Leon. In the 10th century a buffer zone was established by the Christians as the "land of castles," or Castile. Under French leadership, Christian crusades were undertaken against the Moors and strongholds were established along the Pyrenees to maintain a buffer against Islamic Spain, thereby creating the kingdom of Aragon and the county of Catalonia.

In the late ninth century Mozarabs, Christians living in the Muslim controlled southern part of the peninsula, fled to the north as the *Reconquista* pushed southward, and formed a stronghold at Porto (Oporto). The king of Leon reorganized this territory into the province of Portugal.

Thus the medieval coinage of the Iberian peninsula is a combination of Muslim designs, especially in the south, and of increasingly Christian designs as the *Reconquista* progressed. Success of the *Reconquista* became a theme in the designs on Christian coinage of Iberia, and astronomical symbols are abundant, largely because of the influence of Arab astrologers.

Christian coinage in Leon-Castile began with Alfonso VI (1073–1109), who recaptured Toledo in 1085. His dineros contain either a Christogram (a combination of the greek letters *Chi* and *Rho*) or annulets and stars (figure 103). The dineros that contain celestial symbols were probably issued after 1084, when an annular eclipse cut through central Leon-Castile and directly over Toledo (figure 104). This eclipse may have been taken as an omen for the recapture of Toledo.

Figure 103. Dinero of Alfonso VI

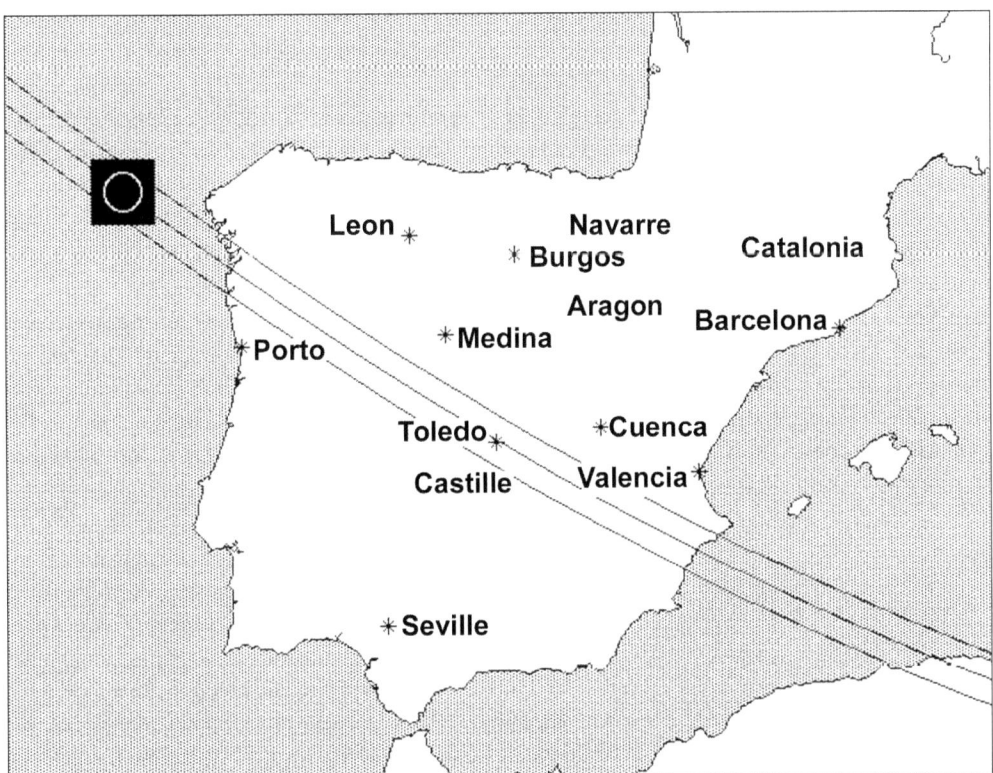

Figure 104. Path of the 1084 eclipse over the Iberian Peninsula

Coinage in Aragon began during the reign of Sancho Ramirez (1063–94). His dineros have a bust on the obverse and a floral cross on the reverse. Some of these dineros contain crescents in the design. These may be interpreted as leaves added to the floral design, or may be crescents that reflect the appearance of the sun as seen from Aragon during a 1079 total eclipse or the 1084 annular eclipse (figure 105). The flower and crescents motif was immobilized in Aragon for the next 100 years. Alfonso I (1104–34) also issued one type of dinero with an annulet on the reverse (figure 106), which may have commemorated the 1084 eclipse and the victory at Toledo.

Figure 105. Dinero of Sancho Ramirez

Figure 106. Annulet type of Alfonso I

Feudal French Provinces

Political and financial agreements between the king and local clergy and nobility in the French feudal provinces led to independent authority to issue coinage beginning in the 10th century. Depending on the local customs and beliefs of each independent authority, astronomical symbols were either absent or profusely used in the coinage design throughout the medieval period.

The main purpose of feudal coinage was to facilitate trade. Acceptance of local coinage was a prerequisite for successful economic regions. Thus, ruling authorities often attempted to maintain economic prosperity by immobilizing accepted coinage designs, or by copying successful coinage of nearby regions. A secondary objective of the designs was propaganda, a way to influence the local population to support the ruling authority or to demonstrate alliance with another province.

The local historical documentation of not only the astronomical events that may have been associated with coinage design, but also of who had local authority and the period of that authority, varies from province to province. If the local authority were associated with a monastery, the records were generally better kept than if a civil authority were predominant.

Analyses must therefore rely on data that may be somewhat incomplete or questionable. Many of the excellent examples of coins with astronomical symbols are so poorly documented that the issuing authority remains anonymous, and attribution has been ascribed to the province based on limited legend or design similarity. Even if the correct province is known, examples exist where several rulers with the same or similar names prohibit attribution with certainty.

Figure 107. Denier of Alain IV

Alain IV (1084–1148), duke of Brittany, changed the design of the reverse cross on the denier to one formed by crescents (figure 107). In 1084, the annular eclipse that crossed the Iberian peninsula was seen as an 86 percent partial eclipse in Brittany (figure 108).

Figure 108. 1084 eclipse view from Brittany

Byzantium

In Constantinople, near the eastern end of the eclipse path, the reverse of a bronze follis of Alexius I (figure 109) issued between 1081 and 1092 shows solar symbols made up of large pellets surrounded by small pellets and a prominent crescent similar in shape to the view of the 1084 eclipse as seen from that location.

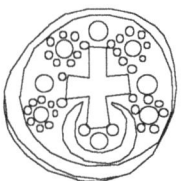

Figure 109. Alexius I

The Total Eclipse of 1133

The total eclipse of 1133 was of moderate duration. The path of totality ranged from 209 to 264 kilometers in width, with a maximum view of totality lasting for 4 minutes and 38 seconds (figure 110). Totality was viewable between 10:33 and 13:26 UT. This eclipse was visually spectacular, and was reported in Ireland, Scotland, England, Holland, France, Germany, Italy, Austria, Hungary, and Byzantium. In *Honorii Augustodensis: Summa Totius et Imagine Mundi* (quoted in Stephenson 1997), the following passage refers to this eclipse as seen in Augsburg, Germany:

> An eclipse of the Sun occurred on the 4th day before the Nones of August at midday for about an hour, such as is not seen in a thousand years. Eventually the whole sky was dark like night, and stars were seen over almost the whole sky. At length the Sun, emerging from the darkness, appeared like a star, afterwards in the form of a new Moon; finally it assumed its original form.

Ireland

One of the Hiberno-Norse pennies issued in Ireland between 1110 and 1150 contains a rayed-annulet in the fourth quadrant of the reverse design (figure 111). Similar Irish coins of this period have annulets and crescents.

46 Astronomical Symbols on Ancient and Medieval Coins

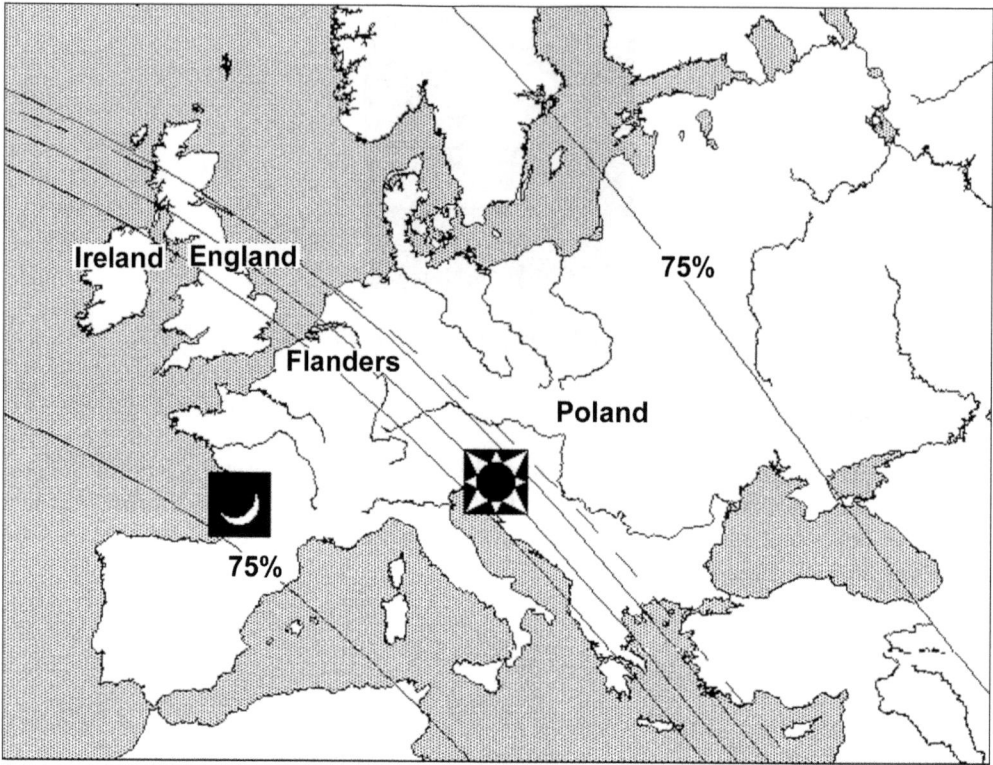

Figure 110. Path of the August 2, 1133, total solar eclipse

Figure 111. Irish penny

England

In nearby Norman England, pennies of Henry I (1100–35) contain stars, and one type (figure 112) may be related to this eclipse. However, it is more likely that this type was struck prior to the 1133 eclipse, and that the star is an immobilized motif. This coin along with the entire Norman series is discussed later.

Figure 112. Henry I of England

Feudal French Provinces

In the French feudal provinces of Flanders and Artois, varieties of petite deniers were struck during the 12th and 13th centuries that contain mullets and other astronomical symbols. Some of these petite deniers cannot be dated with any degree of certainty, while others were probably struck during the first half of the 12th century. The mullets on these coins may represent the 1133 total solar eclipse or one of a different date.

Two petite deniers from St. Omer (figures 113–114) have mullets and may have been struck at about the same time, as both coins use the gothic form of the letter *m* and an epsilon for the letter *e*. The cross fleury of the abbey piece is a typical cross design of the first half of the 12th century.

Figure 113. St. Omer: Dietrich of Alsace (1128–68)   Figure 114. Abbey of St. Omer

Poland

A penny of Wladislaw II struck in Krakow between 1136 and 1146 has a star in the design next to a bust of the king holding a sword (figure 115).

Figure 115. Wladislaw II of Poland

The Annular Eclipse of 1153

The magnificent annular eclipse of 1153 was close to the maximum possible size and duration. The path of annularity ranged from 333 to 403 kilometers in width (figure 116). Annularity occurred between 10:22 and 12:46 UT. At its maximum view, annularity was visible for 6 minutes and 53 seconds. As with the annular eclipse of 1147 (discussed in Appendix B), viewing conditions for this spectacular event must have been quite favorable, as examples of coins with solar symbols exist all along the eclipse path. The eclipse was recorded in France, Germany, Austria, and Italy.

By the middle of the 12th century, the notion of heresy had reappeared in Europe on a scale not experienced since the 5th century. Events of the first half of the 12th century demonstrated a great chasm between religious idealism and actual papal administration. The second half of the 12th century saw religious schism and lack of control by the church. With minimal

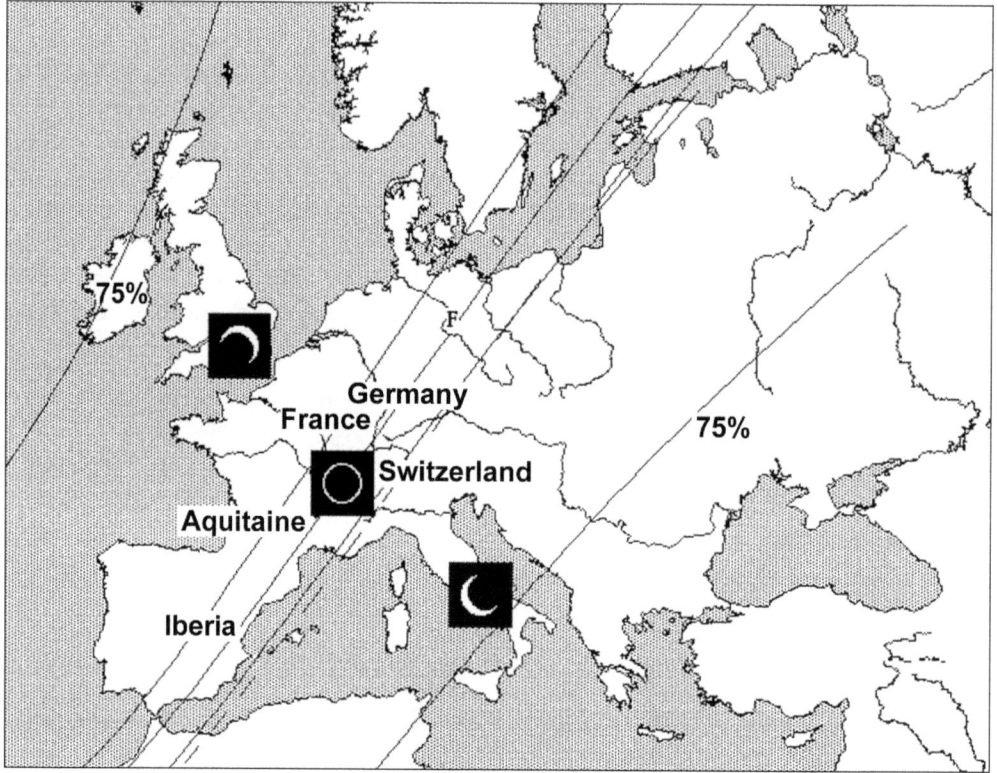

Figure 116. Path of the January 26, 1153, annular solar eclipse

church influence over secular matters, astronomical symbols on European coinage were abundant.

The Iberian Peninsula

The total solar eclipses of 1133 and 1140 and the annular eclipse of 1147 would have been seen as crescents on the Iberian peninsula. However, the 1153 annular eclipse crossed directly over the eastern part of Castile (figure 117), and some of the coins of Alfonso VII have annulets or a combination of stars and annulets. In Leon-Castile, various coins of Alfonso VII (1126–57) were issued with annulets, crescents, and stars (figures 118–120).

In Aragon, Alfonso II (1162–96) struck dineros for use in the feudal French area of Provence, which also was in the eclipse path of annularity, that had an annulet at each side of the base of a Calvary cross (figure 121). Just to the north of the eclipse path in Navarre, Sancho VI (1150–94) struck denaros with crescents and mullets (figure 122).

The annulet became standard in the design of all coinage of the eastern Iberian peninsula beginning in the second half of the 12th century, most likely as a representation of the great annular eclipse that traversed the region in 1153. The non–Aragon or feudal issues of Catalonia, which included Barcelona, land held by the Carolingian kings of France, and lesser bishoprics and municipal leaders, all adopted annulet designs (figures 123–125).

3. Eclipses   49

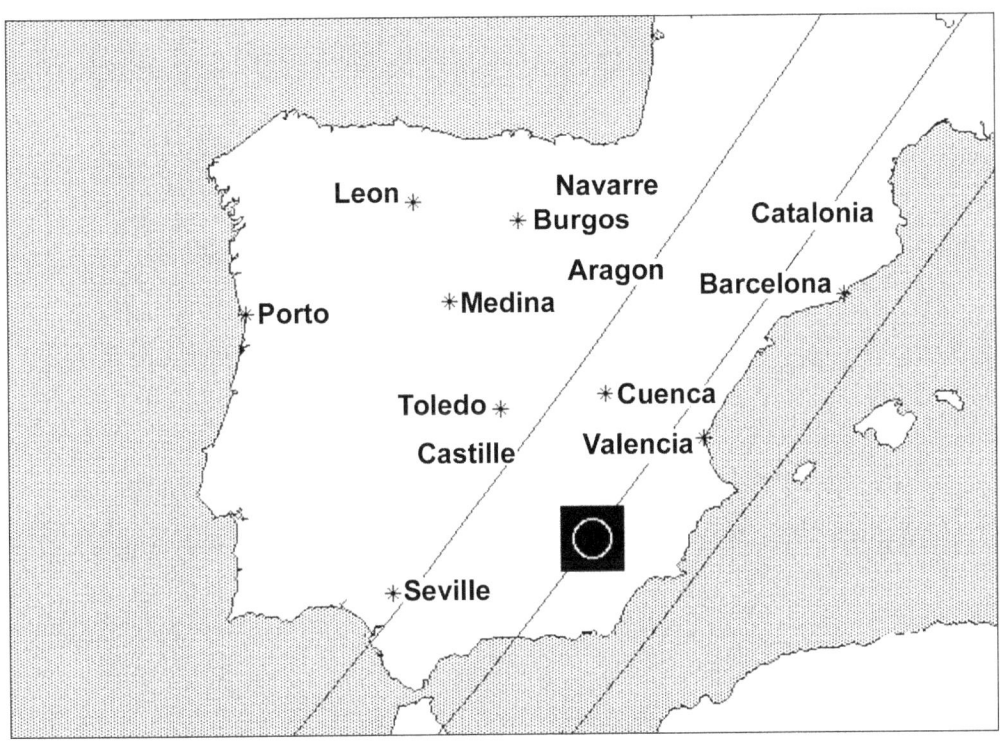

Figure 117. Path of the 1153 eclipse over the Iberian peninsula

Figure 118. Crescent and star type of Alfonso VII

Figure 119. Annulet type of Alfonso VII

Figure 120. Annulet and star type of Alfonso VII

Figure 121. Dinero of Alfonso II

Figure 122. Denaro of Sancho VI

Figure 123. Peter I of Catalonia (1196–1213)   Figure 124. James I of Catalonia (1213–76)

Figure 125. Hugo III/IV (1153–88), counts of the Catalonian Province of Ampuria

Aquitaine

Henry II, duke of Aquitaine (1152–72), issued the first Anglo-Gallic denier, and the reverse has four annulets in the design (figure 126). Ainslie (1830) suggested that these annulets are related to the three annulets on the shields of the kings of Castile, Leon, and Aragon and the counts of Barcelona, and in particular, to Alfonso VIII of Castile, who married Aleonor, the daughter of Henry II. Ainslie discounted the use of the annulet as a symbol of eternity on such a perishable coin as a billon (low silver content) denier, and therefore concluded that the annulets must have been intended as a compliment to the Spanish throne. However, Alfonso VIII was born in 1155 and the denier of Aquitaine may have been struck before his birth. The correlation of the Anglo-Gallic annulletted coinage with Spanish design is correct, but common viewing of the same eclipse is a more likely link.

Figure 126. Henry II denier

Henry II of England acquired Aquitaine through his marriage to Eleanor in 1152 following the annulment of her marriage to Louis VII of France that same year. Early Aquitaine coins of Louis VII do not have astronomical symbols (figure 127), but his later coins are found with either an annulet or a star in the design (figures 128–129). The "+ AQVI TANI E" in four lines across the reverse field of Henry's coins is similar to the later coins of Louis VII with "DVX AQVI TANI E" in four lines.

Thus, the Aquitaine coins of Louis VII without annulets were probably issued before 1153, as the annular eclipse of 1147 did not cross over Aquitaine. As English dukes and kings tended to continue coinage designs in keeping with to local custom, the similarity of Henry's coins to

Figure 127. Louis VII denier of the first type   Figure 128. Louis VII denier of the second type

Figure 129. Louis VII obole

those of Louis VII suggests that perhaps coins were still being issued in Aquitaine in the name of Louis VII after January 1153, and that those of Henry II were issued concurrently or followed sometime later. Louis VII did not relinquish his title of duke of Aquitaine until August 1154, and these coins could have been issued during this 20 month period. Perhaps Louis VII was using the eclipse to assert his right to Aquitaine. Certainly the Louis VII coins were not a compliment to the Spanish throne.

Feudal French Provinces

Many of the deniers of 11th and 12th century Issoudun are characterized by an immobilized design consisting of a gothic *m* with a bar above and an annulet below (figure 130). The annulet may represent the 1033 annular eclipse that crossed the middle of France or an earlier celestial event. In the beginning of the second half of the 12th century, Eudes III, who died around 1164, added crescents to the second and third quadrants of the cross and replaced the annulet with a third crescent (figure 131). The 1153 eclipse would have appeared as a thin crescent in Issoudun (figure 132).

Figure 130. Issoudun denier   Figure 131. Denier of Eudes III

Burgundy was in the path of the 1153 annular eclipse, and deniers of Hugues III (1162–93) have annulets and stars (figure 133). Coins of his immediate predecessors are unknown, but those of Eudes I (1067–1102) also have annulets, and the design may have been immobilized from previous issues.

In the French feudal province of Champagne, the bishop of Meaux, Renaud (1158–61), struck two deniers, one with three stars surrounding his croziers and another where an annulet

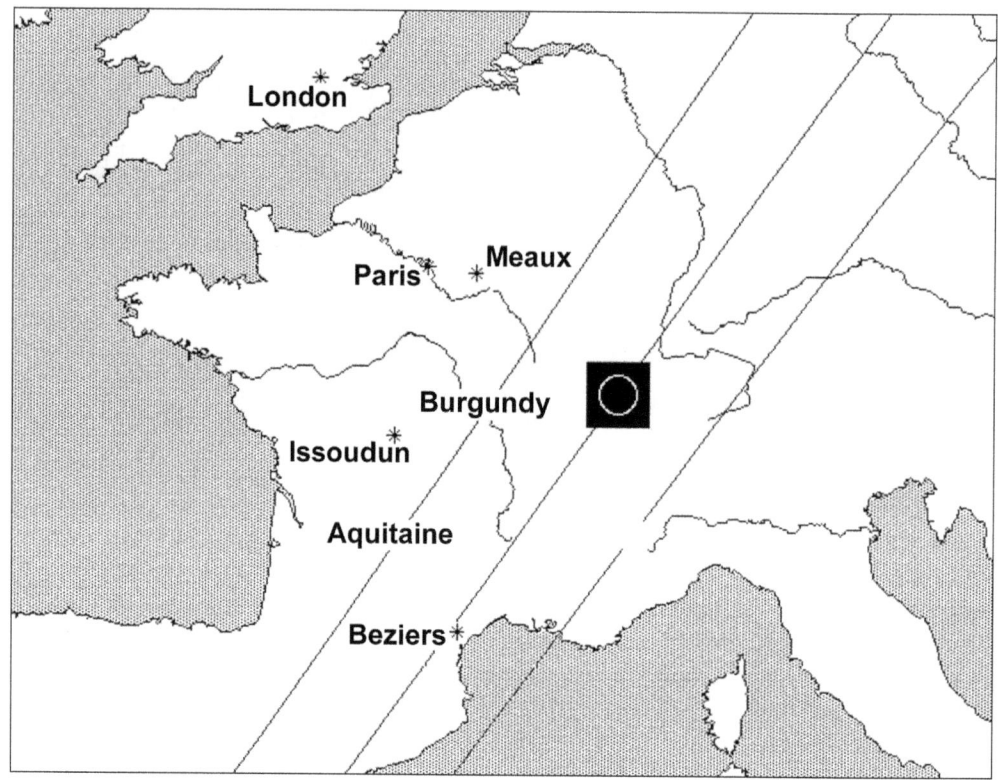

Figure 132. Path of the 1153 eclipse over France

Figure 133. Hughes III denier

replaces one of the stars (figure 134). The 1153 eclipse crossed here as well, and as in Burgundy, coins of Renaud's immediate predecessor Manassé II (1134–58) are unknown. Meaux deniers of Buchard (1120–34) do not contain stars or annulets.

Three deniers of the Languedoc region of Beziers are interesting examples of the use of astronomical symbols as a propaganda tool. Following the death of Roger I in 1150, Bernard-Hatton II, viscount of Nimes, contested the succession of his older brother Raymond-Trencavel. Bernard-Hatton issued two deniers in his name, one with pyramids and annulets and another with mullets in the design (figures 135–136). The pyramids probably represent Halley's comet of 1145 or the comet of 1151. Comets represented the deaths of princes and the changing of kingdoms. The annulets and mullets represent a solar eclipse, another sign of divine right.

Not to be outdone, Raymond-Trencavel also issued a denier with mullets (figure 137). It is possible that the coins were issued after the annular eclipse of 1153 and that the celestial symbols were used to represent the eclipse.

#

Figure 134. Renaud of Meaux   Figure 135. Bernard-Hatton pyramids and annulets

Figure 136. Bernard-Hatton mullets   Figure 137. Raymond-Trencavel mullets

Switzerland and Western Germany

In Switzerland, Friederich I Barbarosa (1152–89) issued a bracteate (coin struck on only one side) with an annulet above a castle gate (figure 138), and in Aachen, a penny with a star above the city (figure 139).

Figure 138. Switzerland   Figure 139. Aachen   Figure 140. Brandenburg

Northern Germany

Brandenburg was directly in the path of annularity, and at some point during his reign, Albert the Bear (1134–70) struck a bracteate with an annulet at the top of the staff (figure 140).

The Total Eclipse of 1241

Europe was treated to three total solar eclipses during the first half of the 13th century, but eclipse symbols are generally absent from royal coins of western Europe of this period. The 1230 eclipse crossed Scandinavia and England, and the 1239 eclipse (discussed in appendix B) was recorded in Portugal, Spain, France, England, Germany, Italy, Austria, and Hungary. By the middle of the 13th century the Albigensian crusades (1209–29) and the subsequent systematic operation of religious inquisitions against heresy between 1230 and 1250 had strengthened the rule of the church over secular matters. Astrological beliefs did not die out in the 13th century, but became more of a pseudo-scientific explanation of questions that the church could not answer. This may explain the lack of astronomical symbols on the royal coinage of France, Italy, and Spain during this period.

In 1241, only two years after the 1239 solar event, the third European total solar eclipse occurred. This eclipse was of moderate duration. The path of totality ranged from 210 to 294 kilometers in width, with a maximum view of totality lasting for 3 minutes and 37 seconds (figure 141). Totality was viewable between 11:24 and 13:24 UT. It was recorded in England, Denmark, Holland, France, Germany, Austria, Italy, and Hungary.

This eclipse must have been spectacular as it crossed highly populated areas and it was symbolized on numerous coins. Despite the strict interpretations of the church, two total solar eclipses in such a short time surely made an impression on the general population. During the second half of the 12th century and into the next one, new mullet designs on coinage were struck and others became immobilized throughout the continent.

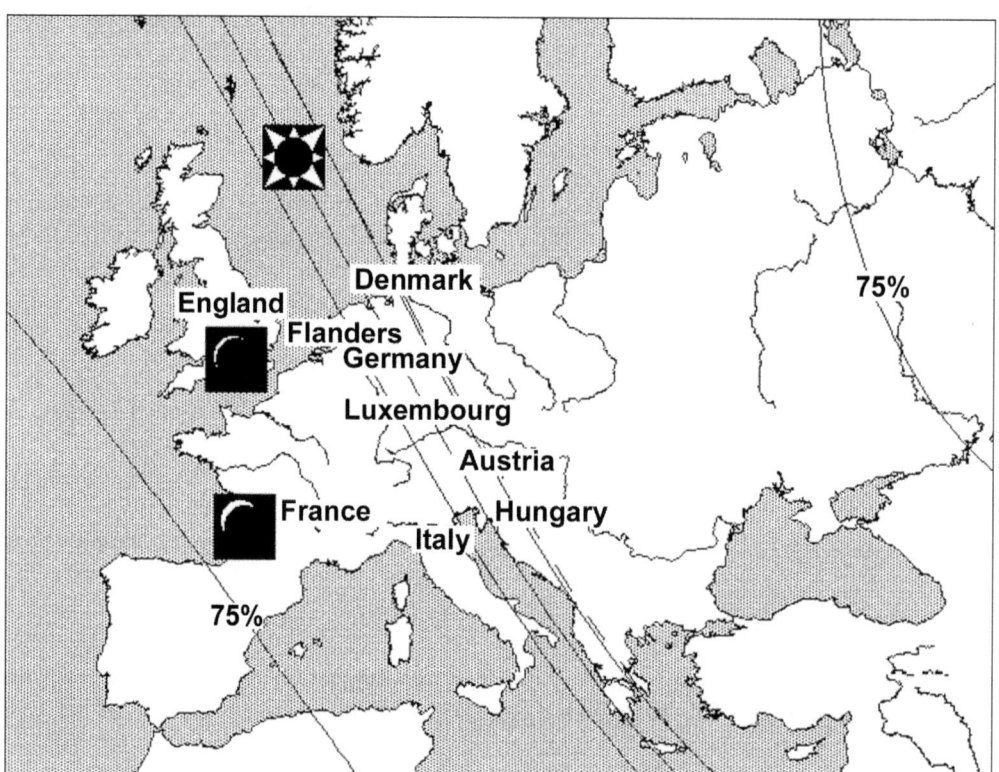

Figure 141. Path of the October 6, 1241, total solar eclipse

England

Some time between 1216 and 1247, Henry III of England struck farthings that have a crescent and pellet in the legend (figure 142). His pennies struck between 1247 and 1250 contain a crescent and star, or only a star, at the beginning of the legend above the king (figure 143). These symbols may be related to the 1241 eclipse or to one in 1230, as both appeared as thin crescents in England. During this period in England, the church was at the height of its power. Perhaps the removal of the astronomical motif after 1250 was the result of pressure from the church.

Henry's Irish pennies struck between 1251 and 1254 do, however, have a mullet next to his bust on the coin (figure 144).

Figure 142. Henry III farthing

Figure 143. Henry III (1247–50) pennies

Figure 144. Irish penny of Henry III

Denmark

Beginning in 1241 with the ploughpenny of Eric (figure 145), a star or star and crescent became the immobilized coinage motif in Denmark for the next 100 years.

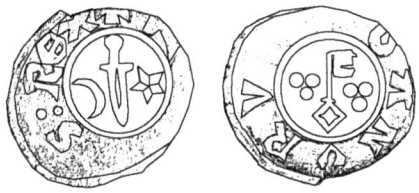

Figure 145. Ploughpenny of Eric (1241–50)

Germany

Coinage of the German Holy Roman Empire, including local coinage in regions such as Holland, Swalenberg, and Lippe, used mullets from the mid–13th until the early 14th century.

Issues of Kaufbeuren (figure 146), probably struck during 1240–50, are found with either a crowned head or two stars above the cathedral. These types are similar to those of neighboring Ravensburg that were issued in the second half of the century, usually without astronomical symbols or with three pellets above the cathedral. One Ravensburg type, however,

56   Astronomical Symbols on Ancient and Medieval Coins

has a star within the doorway of the building (figure 147). The path of the total solar eclipse of 1241 passed directly over Kaufbeuren, and the eclipse would have been seen as a thin crescent in Ravensburg.

Figure 146. Kaufbeuren   Figure 147. Ravensburg

The coinage of William of Holland (figure 148), emperor of the Holy Roman Empire (1247–56), contains mullets, probably in reference to the 1241 eclipse. In nearby Holland, regent issues (figure 149) of Floris V (1256–58) portray stars (or mullets?), and a mullet design was immobilized beginning with his regal coins in 1285 (figure 150). Similar coinage is found in Swalenberg (figure 151).

Figure 148. William of Holland   Figure 149. Floris V as regent

Figure 150. Floris V as king   Figure 151. Vitekind of Swalenberg

In Lippe, pennies (figure 152) issued toward the end of Bernard III's reign (1229–65) were styled after the Irish pennies of Henry III of England (figure 144), and thus were struck no earlier than 1251. A mullet is located next to the bust.

Figure 152. Bernard III of Lippe

Flanders and Luxembourg

Various petit deniers of Flanders, issued in the 13th century contain mullets or stars (figure 153). Luxembourg's Henry IV (1288–1304) struck pennies with mullets (figure 154).

Figure 153. Flanders

Figure 154. Henry IV of Luxemboourg

Austria and Hungary

Duke Bernhard II of Austria (1201–56) issued a pfennig with mullets (figure 155). Bela IV of Hungary (1235–70) issued oboles with mullets representing the total solar eclipses of 1239 and 1241 (figure 156).

Figure 155. Bernhard II of Austria

Figure 156. Bela of Hungary

Italian States

Various designs on coinage of the Italian states of the second half of the 13th century included stars or mullets (figures 157–158).

Figure 157. Pisa

Figure 158. Parma

Feudal French Provinces

In the western feudal provinces of France where the eclipse was seen as a crescent and local customs of using astronomical symbols prevailed, new deniers were issued that portrayed the 1241 eclipse. In Auvergne, the eclipse occurred in the first year of the reign of Alphonse de France (1241–71), and he struck deniers with a crescent and mullet motif (figure 159). Another of his deniers has an annulet replacing the crescent in the quadrant of the cross (figure 160). The annulet may represent the moon or a later eclipse. An annular eclipse crossed Auvergne in 1270 and perhaps this denier was issued toward the end of his reign.

Figure 159. Auvergne mullet and crescent   **Figure 160. Auvergne mullet and annulet**

Elsewhere in the feudal provinces of France, similar mullet and crescent motifs were placed on local coinage (figures 161–162), and such motifs continued to appear through the end of the 13th century.

 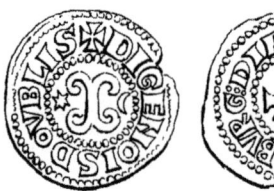

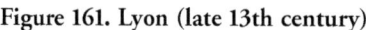 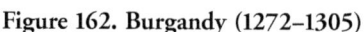

Figure 161. Lyon (late 13th century)   **Figure 162. Burgandy (1272–1305)**

Later Medieval Eclipses

Several late 13th century and early 14th century eclipses are depicted on European coinage, and some of them are discussed in subsequent chapters and in appendix B. The use of astronomical symbols on European coinage was in the early stages of transition from being a source of propaganda to a means of currency control when these eclipses occurred. After Pope Nicholas III died in 1280, Charles of Anjou, who had conquered Sicily and built up alliances among the Italian cities, succeeded in having the subservient Simon de Brie installed as Pope Martin IV. This was the beginning of royalty regaining control from the church. The complexities of Italian politics, combined with failures in the Holy Land, were more than the church could handle.

By the end of the 13th century, astrological symbolism on royal coinage was being renewed in western Europe; however, as nations began to emerge from the darkness of the medieval period in the 14th century, the use of astronomical symbolism on coinage as astrological devices changed to that of marks of mints or indicators of coinage reform. Perhaps such marks were suggested by actual celestial events and were used to make it easier for the general population to identify modifications in coinage weight or metal content. This transformation of astronomical symbols on coins was completed by the second half of the 14th century in most of Europe.

Comets

Denarius of Boleslav II of Bohemia (967–999)

Multi-tailed comet Mrkos

Comets were of particular astrological importance to ancient and medieval man, as they were seen as portents of the deaths of rulers and the changing of kingdoms. One or more comets are usually visible to the naked eye every year or so, although spectacular comets may be seen only once in a generation. Only the truly great comets or those associated with significant events were recorded in historical writings. As the appearance of a comet can take many shapes, depiction on coinage design could also vary greatly, and a variety of ancient and medieval coins portray these celestial visitors. Some numismatists assign different interpretations to cometary symbols on coinage. For example, the symbol below the horse in figure 163 is often described as a lyre, but I will show that celestial symbols are frequently found in this location on such coinage.

Figure 163. Coin of ancient Gaul with a comet under the horse

Brightness, Size, and Shape of Comets

The apparent brightness of a comet is affected by its distance from the sun and the Earth and by the darkness of the sky when it is observed. Some comets have been bright enough to be visible in the presence of the full moon, as well as during daylight.

The apparent length of a comet's tail depends not only on the physical nature of the comet, but also on the geometry of the comet, the Earth, and the sun. Generally the tail will point directly away from the sun. Thus a long tail may appear to be short and bright if all the reflected light from the tail appears to be compacted because of the geometry of observation, and a bright tail may seem dimmer if seen broadside.

Comets are found with a variety of shapes, and a comet can change its shape over a period of a few days. This is due to the physical processes at work in producing cometary tails. Without going into scientific detail as to the physical structure of comets and the tail-producing mechanisms, a brief discussion of cometary shapes is required to understand the possible symbols on medieval coinage.

The head of a comet, or *coma*, is seen as a fuzzy ball, from which both particles and gases are blown away by solar forces. Depending on the mix of particles and gases, the tail, or multiple tails, may be long and narrow (bar shaped), expanding (triangle or spike shaped), or complex and irregular in shape (such as comet Mrkos). The particle tails may also be curved, adding complexity to the shape of the comet. Sometimes, in addition to the primary tail or tails, a thin anti-tail points toward the sun.

Given the different possible combinations of cometary shapes, sizes, and brightness (figures 164–168), ancient and medieval die engravers could have used a variety of different symbols to depict comets on coinage. In addition, the same comet may have been represented differently by various moneyers depending on when the comet was observed and the impression it made on the observer.

Pliny the Elder (Caius Plinius Secundus), head of the western Roman fleet under Vespasian, defined 10 shapes of comets in the first century A.D. in his 37 volumes of *Natural*

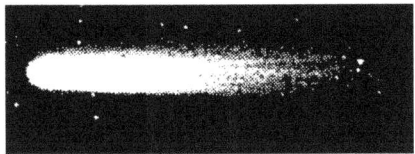

Figure 164. Halley's comet of April 30, 1910 (short bar shape)

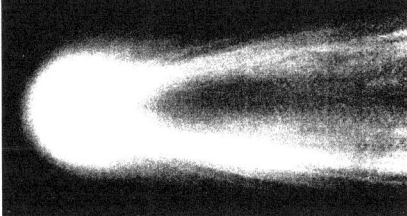

Figure 166. Comet Seki-Lines (triangular shape)

Figure 165. Halley's comet of May 6, 1910 (pyramid shape)

Figure 167. Great comet of 1843 (long bar shape)

Figure 168. 1858 comet of Donati (spike shape)

History. Most ancient coin engravers employed artistic designs to depict cometary shapes, but during the Middle Ages when most coin dies were crudely engraved, comets were represented by variations of simple shapes such as combs, epsilons, pyramids, bars, and stars.

Comets on Ancient Coinage

One of the most famous depictions of a comet on a coin (figure 169) goes back to ancient Rome around 18 B.C., when Augustus Caesar issued a denarius with a comet on the reverse, along with the inscription, "DIVVS IVLIVS" (the divine Julius). This was a direct reference to the comet seen in the heavens immediately after the assassination of Julius Caesar in 44 B.C., and perhaps to the sighting of a comet that was taken as the spirit of Julius Caesar returning to show divine support for Augustus. However, historical records do not report a significant comet at the time this coin was struck, and there is speculation that a bright meteor trail may have been seen. Alternatively, the design may refer only to the comet of 44 B.C.

Comets appeared on ancient coins long before the rise of the Roman Empire. Hazzard (1995)

Figure 169. Denarius of Augustus Caesar

describes a silver tetradrachm of Ptolemy V, who was born in 210 B.C. and became the ruler of Egypt in 205 or 204 B.C. Philip V of Macedon and Antiochus III of Syria immediately launched campaigns to deprive Ptolemy V of his non–Egyptian lands. Territorial losses and native revolts increased, and by 199–198 B.C. the reign of Ptolemy V was in jeopardy. His regent, Aristomenes, needed a way to restore confidence in the young monarch. He struck a tetradrachm as a propaganda piece (figure 170), declaring Ptolemy V as a god manifest in the legend of the coin. The center of the coin depicts the thunderbolt of Zeus flanked by two stars. These two stars represent the comets of 210 and 204 B.C., which appeared at the birth and at the accession of Ptolemy V. Contemporary writings of the period delivered a message of hope: "Although many disasters have plagued the realm, King Ptolemy is a God Manifest, whose greatness was augered in the skies by Zeus for all the world to see!"

The regent saved the dynasty, as the king's subjects believed that celestial events portended an era of prosperity. Thus the depiction of actual celestial events on this coin were clearly used as propaganda.

Figure 170. Ptolemy V

Molnar (1997) suggests that a coin of Mithradates the Great of Pontus was struck for propaganda purposes with a comet as its motif (figure 171). Bright comets appeared in both the year of his birth and in the first year of his reign. These were most likely the comets that were recorded in 134 B.C. and 119 B.C. The comet is depicted on both sides of the coin in the shape of the head and the tail of a horse, presumably referring to the coma and tail of the comet. One of Pliny's descriptions for a comet was *hippeus*, or like a horse's mane in rapid motion. Around 415 A.D. Hephaestion of Thebes wrote in his *Apotelesmatics* that *hippeus* comets

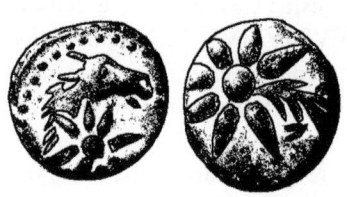

Figure 171. Mithradates the Great

portend the "quick fall of kings and tyrants and rapid changes in the affairs of these countries." Mithradates may have used these comets as omens for the defeat of the Romans in Asia Minor.

Comets on Medieval Coinage

Current numismatic descriptions of medieval coins rarely even suggest that comets were portrayed on the coins. Why would such spectacular and mystical events be included in other medieval records but be excluded from coinage design? The following discussion will show that comets were indeed represented on medieval coins.

European and oriental chronicles recorded a number of comets during the middle ages. For example, the crusaders took the sword-tailed comet of 1097 as a sign of their upcoming victory as they marched into Syria. Similarly, the English populace associated a great comet in July 1198 with the death of King Richard the Lionheart.

One of the most significant comets in history is Halley's (named after Sir Edmund Halley), which returns once every 76 to 77 years. Halley's comet lends itself to good analysis because of the predictability of its appearances. During its 837 A.D. visit, its tail was estimated to be between 93° and 100° in length (from the horizon to the zenith), and astronomers estimate that it was the 12th most brilliant comet ever observed. Of interest in relation to the later Middle Ages is the return of Halley's comet in 1066, 1145, 1222, 1301, and 1456. In Europe the 1066 comet was visible to the naked eye from April 1 through June 7. Although reported to have a short tail of only 3°, it must have been extremely bright, as modern day calculations assign it as the fourth most luminous comet on record. The 1066 comet was regarded as an omen for the conquest of England by William the Conqueror at the Battle of Hastings on October 14, 1066. This comet was described in poetry of the day, and was also shown on the famous Bayeux tapestry (figure 172), where the comet is associated with the death of Harold of England.

The 1145 passage (April 14–July 9) is associated with a strong earthquake and an expedition

Figure 172. Section of the Bayeux tapestry: the comet and Harold's demise

of the king of France. It was pictured in a book of psalms by Eadwine, a Canterbury monk, as having a great tail (figure 173). The 1222 return of Halley's comet was described by chroniclers as going from high in the sky to near the ground, forming a sharp cone. It was visible from September 3 through October 23 and was said to have joined itself with the moon when the moon died. A total lunar eclipse occurred on October 22. As a lunar eclipse occurs when the moon is full, a dim comet would not be seen as it faded in brilliance after the lunar eclipse. This passage of the comet was seen as having heralded the death of Philip Augustus of France.

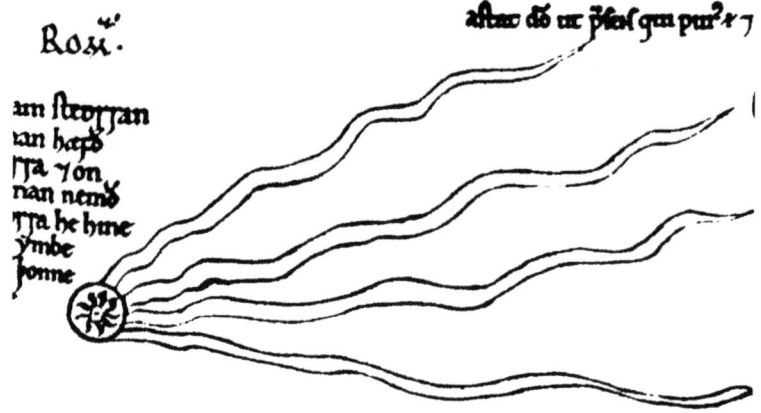

Figure 173. Eadwine's depiction of Halley's comet of 1145

The long-tailed 1301 comet was observed from mid–September through the end of October. It was said to be "full of poetry, but lacked realism," and was given the name *cometa horrendae magnitudinis*. Giotto di Bondone celebrated the comet in his 1303 fresco *The Adoration of the Magi*, in which he depicted the comet as the Star of Bethlehem (figure 174).

Figure 174. Giotto di Bondone's Adoration of the Magi fresco

In 1456 the comet reached from the horizon to the zenith and according to astrologers, was supposed to indicate the success of Mahomet II, who had already taken Constantinople and threatened the rest of the Christian world. In response, Pope Calixtus III ordered extra *Ave Maria* to be repeated, church bells to be rung at noon (the origin of the current custom), and added the prayer: "Lord, save us from the devil, the Turk, and the comet."

Some evidence indicates that comets were believed to be a single star that would reappear. *The Anglo-Saxon Chronicle* recorded several comet appearances as follows:

> [729] In this year the star called "comet" appeared and St. Egbert died...
>
> [975] In this year Edgar's son Edward succeeded to the kingdom. And soon in the same year in harvest time there appeared the star "comet," and in the next year there came a very great famine and very manifold disturbances throughout England...
>
> [995] In this year the star "comet" appeared, and Archbishop Sigeric died.

The Comb as a Cometary Symbol

Consider a series of deniers struck at several mints in the French feudal province of Champagne from the mid–11th century through the early part of the 14th century. An anonymous denier of Sens, and those copied in Provins by Thibaut II (1125–52) and his successors, have combinations of annulets, stars, and crescents above a comb (figure 175). Current numismatic thinking is that the comb (*peigne*) in the central part, or field (*champ*), of the coin is a play on the name of the province. In fact, numismatic references have called this major symbol a comb for many years.

A different explanation, and one that better reflects medieval symbolism, may be correct. Except for the occasional letters in the design field, all the other symbols besides the comb have astronomical counterparts, so why would the comb be any different? A play on words is unlikely to have been meaningful for successive sovereigns over a period of almost 300 years. Also, there is no evidence that the central part of a coin was referred to as a field 1,000 years ago.

Figure 175. Peigne Champenois designs

The Latin word for comet is *cometes*, which comes from the root *coma* and means head of hair. The Latin word for long hair (hairy) is *comatus*, and the Latin word *comans* can be used interchangeably for comet and long hair (hairy). Indeed, this word relationship between "comet" and "long hair" was noted in *The Anglo-Saxon Chronicle*, in reference to the comets of 892 and 1066:

> [892] And the same year after Easter, at the Rogation days or before, there appeared the star which is called in Latin *cometa*. Some men say that it is in English the long-haired star, for there shines a long ray from it, sometimes on one side, sometimes on every side ...

[1066] Then over all England there was seen a sign in the skies such as had never been seen before. Some said it was the star "comet" [Halley's] which some called the long-haired star, and it first appeared on the eve of the Greater Litany, that is 24 April, and so shone all the week.

Comets can sometimes be seen having many tails. When such a comet is viewed close to head on, much of the light reflected from the comet can appear compressed into a bar-like shape with tails or a comb shape. Consider the 1986 return of Halley's comet, when observers saw as many as seven or eight tails (figure 176).

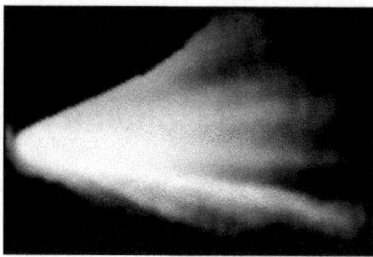

Figure 176. Halley's comet in 1986

The tines of the comb on the earliest of the mid–11th century deniers of Sens are curved rather than straight, and thus are more suggestive of the tail of a comet than a device for grooming hair. The similarity between the numismatic comb motif and the depiction of Halley's comet of 1066 on the Bayeaux tapestry is remarkable (figure 177). The physical similarity and semantic relationship in Latin between multi-tailed comets, long hair, and combs suggests that the immobilized design on this series of French deniers stems from this spectacular comet rather than from a play on words.

Figure 177. Halley's comet and a denier of Sens

The Champagne deniers may not have been the first coins to depict a comet as a comb. In the East Anglia region of Anglo-Saxon England, a gold runic shilling struck around 670 has a star on the obverse and a comb on the reverse (figure 178). Perhaps both symbols represent one of several comets that were visible between 660 and 680, or the star and comb may represent the head and the tail of a comet, respectively.

Other comets may also have been represented by combs on coins. In 1301 Halley's comet

Figure 178. Runic shilling

was described as having a broad tail. At least one French feudal issue probably depicts this comet. In La Marche, Gui (1303) issued a denier with a four-tailed bar in the second quadrant of the cross (figure 179).

Figure 179. Denier of La Marche

In Spanish Majorca, Peter IV (1343–87) issued a variety of denominations with comet-like combs (figure 180). Some numismatists refer to these symbols as scallop shells, but the elongated appearance is not shell shaped. Beginning in 1345, European chroniclers recorded eight bright comets during Peter IV's reign. Earlier issues of Majorca contain mullets in the same locations as the combs (figure 181).

Figure 180. Majorca double real   Figure 181. Sancho of Majorca (1311–24)

The Epsilon as a Cometary Symbol

Many examples of medieval coinage contain an isolated and unexplained epsilon (or variations of *E* shapes) that may represent a three-tailed comet. In this section, epsilon refers to any symbol of this shape. An early example of an epsilon is depicted on a Bohemian denarius of Boleslav II (967–999) with pellets in all but the second quadrant of a cross (figure 182). That quadrant contains a straight-line variety of an epsilon. Two bright comets, those of 975 and 995, were recorded during Boleslav's reign. This symbol has a remarkable likeness to the shape of the 20th century comet Mrkos (figure 183). Perhaps one of the two recorded comets also had three tails.

In the Brittany district of Penthievre, Eudon (1034–79) issued a denier with stars in the

Figure 182. Denarius of Boleslav II   Figure 183. Comet Mrkos

field design and an annulet in the legend (figure 184). The annular eclipses of 1033 and 1044 crossed just to the south of Brittany, as did the annular-total eclipse of 1039. Interestingly, the design includes three stars that may represent these eclipses. There is also an epsilon in the second quadrant of the reverse cross. Note that the epsilon is of a different form than the *E*'s in the legend. Poey d'Avant (1858–62) expressed concern about the attribution of this coin. He assumed that the epsilon is an initial for Eudon, but questions the placement of the letter where a star is usually found. Several comets were recorded during this period including Halley's comet of 1066.

Figure 184. Denier of Eudon

Perhaps the epsilon is an initial for Eudon, but if so, why would it be of different style than the *E*'s in the legend? In Britanny, Pierre Mauclerc (1213–37) struck a denier with an epsilon with an extended middle bar (figure 185). In this case the epsilon cannot be an initial, but perhaps represents an early 13th century comet such as Halley's of 1222.

Figure 185. Pierre Mauclerc   **Figure 186. Denier of Robert II**

In Burgundy, Robert II (1272–1305) replaced the star and crescent on one of his deniers (figure 186, and see figure 162) with symbols similar to those Mauclerc used. This symbol may represent one of several late 13th century comets or the 1301 return of Halley's, which was said to be of unusual shape. Once again, the epsilon cannot be an initial.

During the same period in Aquitaine, Edward I–III of England struck a series of deniers with epsilons (figure 187). The first type was most likely issued in 1291 (see chapter 8 for a detailed discussion). An epsilon is found both under the "AGL" on the obverse and in the first quadrant of the reverse. As with the denier of Eudon, the symbol takes a different form than the *E*'s in the legend. The epsilons may represent several comets that were recorded between 1264 and 1305.

An examination of similar designs on the coinage of Aquitaine supports the hypothesis that the epsilon on these coins is an astronomical symbol. Edward III (1327–77), issued a similar denier with a mullet in place of the epsilon (figure 188). His English and Irish coins

Figure 187. Denier of Aquitaine

that were issued after 1335 have a star in the legend and probably referred to the total solar eclipse of 1330.

Figure 188. Denier of Edward III

Elias (1984) referred to the mullet symbol as a rosette, but on a similar coin issued in the French province of Orange by Raymond III (1335–40), the symbol below the leopard and "AVRE" is clearly a mullet (figure 189). Furthermore, another denier of Edward III replaces the mullet with three pellets in a triangular arrangement (figure 190). Four pellets are also found below the leopard on some of his deniers. On July 3, 1327, in the first year of Edward III's reign, Mercury, Mars, and Jupiter set in the western sky in a 1° triangular conjunction, and Venus was not far away (figure 191). Thus once again, astronomical symbols that may represent an actual event are found in the same position on the coin as the epsilon and support the hypothesis of the epsilon as another celestial symbol.

Figure 189. Denier of Raymond III   **Figure 190. Denier with three pellets**

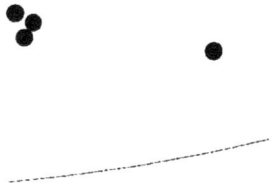

Figure 191. July 3, 1327

Some of the Hiberno-Norse coins of the early 11th century depict a rotated epsilon on the neck of the bust and in one of the reverse quadrants (figure 192). An annulet is found to the right of the bust. This epsilon is different from the hand of God symbol found on some of the Hiberno-Norse issues. Similar coinage issued after about 1060 has three pellets on the bust (figure 193), suggesting that the neck may have been a location for an astronomical symbol. Several comets, annular solar eclipses, and planetary conjunctions can be associated with these coins.

Thus in some cases, the epsilon and its variations may really be a version of a comb symbol used to represent a significant cometary appearance, perhaps one with three tails.

Figure 192. Hiberno-Norse penny with epsilon on neck

Figure 193. Hiberno-Norse penny with three pellets on neck

The Pyramid as a Cometary Symbol

A similarly strong case can be made that open pyramids (*V* shapes) and closed pyramids (triangles) could represent two-tailed comets on medieval coinage designs. Pyramids may be in the form of a plain *V*, a curvy *V*, a plain or curvy *V* with one or three pellets or annulets at the vertex or at each end, or a curved *V* shape similar to a small Greek letter delta. Some numismatists refer to pyramids as piles. Indeed, pyramids and triangles are so prevalent on medieval coinage that these symbols may have been used as generic symbols for any comet with a broad tail.

The comet of 44 B.C. may have been depicted on Celtic issues. The reverse of the Iceni stater of 45–40 B.C. contains a delta or pyramid shaped feature enclosing a pellet that is located above a horse and a solar symbol below (figure 194). Some numismatic sources refer to this as a crude representation of a rider upon the horse. The Celtic horse design is found on other coins with astronomical symbols above and below the horse, and thus one must consider that the pyramid shape is also an astronomical symbol.

Figure 194. Iceni stater

The concept that the pyramid symbol is a rider can be dismissed by examining a Celtic silver unit struck between 35 and 25 B.C. (figure 195). On this piece, the same pyramid with a pellet at its head is found inverted and below the horse with a solar wheel above. Unless the rider fell off his horse, this symbol is more likely the representation of a comet. A third coin struck during the same period replaces the annuletted pyramid with a single-tailed annulet with rays (figure 196). There can be little doubt that it portrays a comet.

Furthermore, immobilized designs of Celtic Trinovantes use the pyramid motif as a comet.

Figure 195. Iceni unit

Figure 196. Comet symbol

On a gold stater that was issued between 45 and 40 B.C., the comet of 44 B.C. is probably represented (figure 197). On the obverse, two crescents with straight decorative rays are found. Two annulets with central pellets and curved decorative rays are also present, as well as two pyramids with a pellet at each head. Comets are often seen with long curved or straight tails. This stater design was immobilized, and one struck between 40 and 30 B.C. retains a similar obverse motif, but on the reverse, a comet is found below the horse (figure 198). Some refer to the figure below the horse on this coin as a cornucopia, but it is more likely a comet.

Figure 197. Trinovantes stater Figure 198. Trinovantes stater

The 12th century deniers of Blois and Vendome are thought to be crude versions of the portrait type of Chinon (figure 199). Some of the types are fairly recognizable as portraits (figure 200), while others are severely distorted, often using astronomical symbols in the design. One of the Blois types replaces the ear pellet with a comet-like pyramid (figure 201).

Figure 199. Chinon denier

Figure 200. Degenerated bust Figure 201. Blois denier

Many of these deniers contain letters, crosses, and astronomical symbols before and after the crude bust (figure 202), suggesting that the pyramid may be astronomical in nature as well.

On some of the deniers, the pyramid has three legs relating the pyramid to the epsilon, and all these pyramid motifs have a comet-like appearance. Furthermore, a few of the degenerated bust deniers have epsilons in place of other astronomical symbols. One of the epsilons is similar to the epsilon types of Brittany and Burgundy, and the other is similar to that found on deniers of Penthievre and Aquitaine (figure 203).

Many medieval coins depict pyramids. Feudal French deniers of Anselme (figure B5) and Beziers (figure 135) may depict Halley's comet of 1145 or a comet seen in 1151. In royal France,

Figure 202. Various symbols

Figure 203. Epsilons

a denier of King Lothaire (954–986) that was issued in Calais has a pyramid in one quadrant of the reverse cross (figure 204). The comet of 975, with a 60° tail, was the only comet recorded in Europe during his reign. A tail of this length would extend two-thirds of the way from the horizon to the zenith.

Figure 204. Denier of Calais

During the first half of the medieval period triangles were used more frequently than pyramids to represent comets, and the symbols can be related to each other. An example can be found in the early 11th century issues of the French feudal province of Berry. Lords of Brosse issued similar deniers, except that triangles surrounded by a solid circle were used in one design (figure 205) and pyramids surrounded by a beaded circle in the other (figure 206). The comet of 995 was seen for 80 days in France as is probably reflected on these issues.

Figure 205. Triangle denier of Brosse   Figure 206. Pyramid denier of Brosse

In 11th century Reims, a denier has triangles in the third and fourth quadrants of the reverse cross, a star in the first quadrant, and a pellet and crescent in the second quadrant (figure

207). Surely the star and crescent are astronomical symbols, and thus the pellet and the triangles may also represent heavenly events.

Figure 207. Denier of Reims

A late issue of Charlemagne (768–814) struck between 793 and 812 in Calais contains triangles in each quadrant of the reverse cross, surrounded by a solid circle (figure 208). This reverse design is very similar to that of the triangle type of Brosse, suggesting that the issue of Brosse may have been struck much earlier than the 11th century.

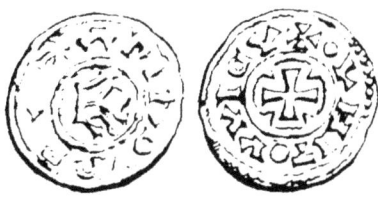

Figure 208. Triangle denier of Charlemagne

No comets were recorded in medieval European chronicles during the time the coin of Charlemagne was struck. The previous documented comet in Europe was that of 676, and no others were recorded until 817. However, a number of comets were recorded in eastern journals during this period.

Charlemagne's coin was struck near Calais, across the Channel from the Anglo-Saxon kingdoms of England. In Kent, Cuthred (798–807) added triangles to his portrait and nonportrait pennies (figure 209). In Mercia, pennies (c. 805) of Coenwulf (796–821) are found with triangles in the quadrants of the cross (figure 210). In Benevento (Italy), a denaro of Grimoald IV (806–817) contains an elongated object in each of the quadrants of the reverse cross (figure 211), and a triangle is retained by the bust of his successor, Sico (817–832). On Sico's coin (figure 212) a crescent is found on the other side of his bust supporting the concept that the triangle was used as an astronomical symbol. One or more late eighth century or early ninth century comets were likely depicted on each of these issues.

In addition to the 1066 return of Halley's comet, that of 837 shares honors for the greatest

Figure 209. Cuthred   Figure 210. Coenwulf

Figure 211. Grimoald IV

Figure 212. Sico

comet of the medieval period. On April 13th its tail was more than 90° in length, that is, it stretched from the horizon to past the zenith. This visit of Halley's comet was the closest recorded approach to Earth in history. This visitation of the comet is sometimes referred to as the comet of Louis the Debonnaire (also known as Louis the Pious, 814–840), son of Charlemagne. Although the comet appeared near the end of his reign, several of the deniers of Louis the Pious may refer to this magnificent celestial visitor. Triangles are found on at least three different types of his deniers (figure 213), and a star is found on a fourth type (figure 214). A total solar eclipse crossed France on May 5, 840, and this may be the source of the star.

Figure 213. Louis the Pious (triangles)

Figure 214. Louis the Pious (star)

In Anglo-Saxon England triangles that may represent this great comet were introduced on coinage. In Mercia, Berhtwulf (840–852) issued a penny with elongated triangles in the quadrants of the cross (figure 215). Sometime between 827 and 845, Aethelstan I of East Anglia replaced pellets with elongated triangles (figure 216). In Wessex, Ecgberht (802–839) struck pennies with crescents or triangles in the reverse design (figure 217). Six solar eclipses between 807 and 818 would have been seen as a thin crescent in England, with the solar eclipse of 809 being recorded in the *Anglo-Saxon Chronicle*. The penny with triangles in the quadrants of the cross most likely represents the great comet of 837. Some of the early coins of his successor, Aethelwulf (839–48), also contain triangles as an immobilized design (figure 218–219), suggesting that Ecgberht's triangle type was struck late in his reign.

In 868 a comet was well documented in eastern chronicles. Aethelred, Archbishop of Canterbury (870–889) issued a penny with triangles in the quadrants of the cross (figure 220) that may represent this celestial visitor.

Figure 215. Berhtwulf

Figure 216. Aethelstan I

Figure 217. Ecgberht

Figure 218. Early penny of Aethelwulf   Figure 219. Late penny of Aethelwulf

Figure 220. Aethelred

The *Anglo-Saxon Chronicle* described the comet of 892: "[892] Some men say that it is in English the long-haired star, for there shines a long ray from it, sometimes on one side, sometimes on every side."

In Canterbury, Aethelred's successor, Plegmund (890–923), issued a penny with two triangles, two annulets, and an annuletted cross on the reverse (figure 221). On August 8, 891, an annular solar eclipse crossed just to the south of England. Perhaps this coin depicted both celestial events. In Carolingian France a denier of Eudes (887–898) has a narrow triangle in the first quadrant of the reverse cross (figure 222).

Figure 221. Plegmund   Figure 222. Denier of Eudes

Medieval Hungarian coinage began with that of Stephen and is characterized by a cross with a triangle in each of the four quadrants on both the obverse and reverse of the coins (figure 223). This design remained as an immobilized type for almost two centuries. Given the revered status of Stephen the immobilization of his design type is understandable. Stephen was born in 975 and assumed power in 997 following the death of Prince Geza. The first issue of Stephen's coins are usually associated with his coronation in 1000.

Figure 223. Denier of Stephen (1000–38)

Western chronicles recorded comets in both 975 and 995. In a manner similar to the previously discussed example of Mithradates the Great, the correlation of cometary appearances with the birth year of Stephen and as an omen of the death of Geza made for a remarkable propagandistic opportunity. Coupled with the recorded knowledge of Stephen's efforts to consolidate power, eradicate paganism, and convert his people, that he used cometary symbols to reinforce his divine right to the kingdom is a natural conclusion.

The Bar as a Cometary Symbol

Single-tailed, nontriangular-shaped comets may have been represented on coins by short and long bars (spikes). Bars on medieval coins were sometimes used as a mark to indicate abbreviation, and as with a single pellet, caution must be taken that an isolated bar is not merely a minor part of the design. However, the use of a bar where other astronomical symbols are found is a good indication that it may represent a comet.

In Byzantium, Basil II (976–c.989) struck a series of gold histamenons. One type has the nimbus bust of Christ on the obverse, and facing busts of Basil and Constantine holding a patriarchal cross between them. Varieties of this type have the obverse nimbus decorated with pellets, bars, triangles (sometimes referred to as wedges on these coins), and pellets within annulets. The reverse staff is found without any symbols (figure 224) or with one or five pellets, a triangle, or a bar near the base (figure 225). Both the triangle and bar may have been used as astronomical symbols on these coins. The comet of 975 may have been taken as an omen of the death of John I Tzimisces in January 976 and the triangle and bar may represent this celestial visitor. In Cilician Armenia, trams were struck with various astronomical symbols on the base of the staff between Hetoum and Zabel (1226–71), including a star, supporting the hypothesis that astronomical symbols adorned staffs in the eastern regions.

Halley's comet of 1066 was spectacularly bright and had a short tail. The comb deniers of Sens depicted this comet as a "hairy" bar. In Hungary, Geza I, as duke from 1064–74, and as king from 1074–77, replaced part of the immobilized triangle design of Stephen with bars (figure 226). The annulets on Geza's coin were immobilized from earlier coinage, perhaps representing a solar eclipse in 1059.

In Hungary, several of the coins of Koloman (1095–1114) have crescents, pellets, triangles, and bars in one quadrant of a cross. Although these symbols may have been inspired by actual celestial events, they may be marks of moneyers or mints because there are many variations and combinations of symbols used on the same type of coin. Bars, however, are a clearly a major

Figure 224. Plain staff

Figure 225. Bar on staff

Figure 226. Geza I

element on one of his deniers (figure 227). During Koloman's reign, comets were recorded in European chronicles in 1097, 1106, 1109, 1110, and 1114. The tail of the comet of 1114 extended over a great distance, and therefore a long bar may have been used to represent it. Design transition evidence suggests that this bar type may represent the comet of 1114 rather than one of the earlier events. Stephen II (1114–31) issued a similar coin (figure 228), but with the immobilized triangles of Stephen I replacing the bars, thereby providing a transition of the design from the end of Koloman's reign to the beginning of Stephen's.

Figure 227. Koloman   Figure 228. Stephen II

The 1222 return of Halley's comet was described as going from high in the sky to near the ground, forming a sharp cone. In Portugal, mealhas of Alfonso II (1211–23) contain triangles, and the early dineros of Sancho II (1223–48) contain spikes, or sharp triangles (figure 229). There can be little doubt that the spikes in the dineros of Sancho II represent the sharp cone of this return of Halley's comet, and that the comet would have been taken as a sign for his succession.

Figure 229. Dinero of Sancho II

The 1222 passage of the comet was seen as a portent of the death of Philip Augustus of France. In the French feudal province of Nevers, Gui de Forez of Nevers (1226–41), issued a denier with a comet symbol that included both a head and a tail (figure 230).

Figure 230. Denier of Nevers

In Cilician Armenia, Hethoum I (1226–71) issued a large bronze tank with stars and long bars in the quadrants of the reverse cross (figure 231). This coin was similar in design to the Christian coinage of western Europe.

Figure 231. Cilicia

An interesting variation of the bar representation of a comet is found on a unique penny of Cnut, struck at the Norwich mint in Anglo-Saxon England (c. 1017–23). Cnut pennies of this type are sometimes found with symbols such as crosses or pellets near the bust or on the reverse. This coin depicts a pointed bar with a pellet (figure 232). Originally described in the numismatic literature as a dagger, it more likely represents a comet. The Danish invasions of England saw much bloodshed, but by the time this coin was struck Cnut had conquered his opponents, and England remained in a fairly peaceful state during the rest of his reign.

Figure 232. Cnut penny with comet symbol

Cnut's father, King Swein, had invaded England in 1013. Upon Swein's death on February 3, 1014, the Danish fleet proclaimed Cnut king. During 1014–16 England was also claimed by the Saxon king, Aethelred II, and then by his son. During 1015 and 1016 Cnut returned to do battle, and finally secured the title of king of all England in November 1016 after both Aethelred

II and his son had died. Two years later, in 1018, Cnut inherited the Danish kingdom when his brother Harold died.

During this period medieval astronomers recorded three comets. From mid–February, one week after the death of Swein, through March 1014, oriental astronomers recorded a comet with a long tail that would have been visible in England if clear skies had prevailed. In 1017 European astronomers documented a comet with a large tail that lasted for four months, and in August 1018, both European and oriental astronomers wrote of a comet with a tail that stretched across a third of the sky. The 1018 comet was visible for 30 successive nights in Northumbria (northern England) and was seen as a harbinger of a future disaster (Fletcher 2003).

Without additional information, whether this coin symbol represents a comet or a dagger is unclear, but the cross-guard of a dagger is not part of the symbol. Pliny the Elder had described one of the possible comet types as a dagger shape (*xiphias*) as early as the first century. Given the peaceful state of England during 1017–23 and the fact that the daggers shown on Viking coins of England a century earlier always had cross-guards, a dagger-shaped comet is the more likely representation. A comet depicted in Johannes Hevelius's *Cometographia* (1668) is similar to the symbol on this coin (figure 233).

Figure 233. Comet type from Cometographia

Under the assumption that the symbol is a comet, then which comet is represented? The long-tailed comet of 1017, and those of 1019 and 1020 (neither of which was recorded in Europe), were not associated with a change in sovereignty. Both the comets of 1014 and 1018 appeared at a time when Cnut assumed a new kingdom. The comet of 1014, however, was not recorded in Europe, but the comet of 1018 was, and appeared during the time that this type of coin of Cnut was struck.

The Star as a Cometary Symbol

During the medieval period in Europe, comets were infrequently depicted as stars on coinage, perhaps to better distinguish them from solar events. However, in a few instances stars may have been used as cometary symbols, especially if the star symbol has a tail.

In the feudal French region of Languedoc, bishops of Mende struck a denier with a star bearing a tail (figure 234). Although the dates for this coin are not known, the style is that of the 13th to early 14th century.

Several comets were documented in Europe during the 14th century. One or more of these comets may have been represented on some of the coinage of northern German towns. In Wismar, a star within an annulet was added after 1380 to the reverse design of its local coinage (figure 235). Comets were observed in 1378 and 1382, and may be depicted on the coin.

In England, pennies of Edward the Confessor demonstrate an interesting transition between stars and pyramids as possible cometary symbols. The facing bust type, struck around

Figure 234. Bishops of Mende Figure 235. Wismar

1062–65, has a small cross on the reverse. One rare variety struck in Sudbury depicts the cross as a star (figure 236).

Figure 236. Star cross type

Edward's next penny, struck around 1065–66, has a profile bust and pyramids on the reverse (figure 237). A rare penny struck at a few mints combined the facing bust obverse with a pyramid reverse as a transitional type (figure 238). Although no comets were recorded in Europe when these pennies were issued, eastern chronicles recorded three comets in 1063 and one in 1065.

Figure 237. Pyramid type Figure 238. Transitional type

The Confessor was ill during the last years of his reign, and perhaps one of these comets was taken as a portent of a change in kingship. However, the star cross type of Edward may depict the annular solar eclipse in 1064 that crossed just to the north of Scotland and was seen as a thin crescent in England. Facing bust types exist with a crescent in the reverse field (figure 239) and with

Figure 239. Cross and crescent Figure 240. Sven Estrithson

an immobilized annulet from the York mint. In Lund, Denmark (now Sweden), where coin types were similar to those of Anglo-Saxon England, Sven Estrithson (1047–74) struck a penny later in his reign with both an annulet and crescent in the reverse field (figure 240). The eclipse was seen as a thin crescent there as well.

Comets on Post-Medieval Coinage

The appearance of a comet, especially a bright one or one with a long tail, continued to mystify mankind and generate fear throughout the Middle Ages and into the post-medieval period. Although comets were no longer a portent of changing kingdoms after the medieval period, they were still depicted on contemporary coinage, and usually as a star with a long tail (figure 241).

Figure 241. German ducat showing the comet of 1618

Complex and Unusual Astronomical Designs

Julian II (360–363): Venus and Mars in Taurus

5. Complex and Unusual Designs

Previous chapters showed how symbols for the sun, moon, planets, stars, eclipses, and comets represented deities, heavenly objects, and actual celestial events. When a number of celestial events took place, when interplay occurred between multiple types of astronomical objects, or when celestial events could be combined with other forms of iconography, complex designs were used. These types of designs can be found on many forms of ancient and medieval artwork as well as on coinage.

For example, the tomb of Antiochus I, ruler of ancient Commagene, depicts a lion with three large stars above it, 19 smaller stars on it and in the background, and a crescent moon on its neck (figure 242). Neugebauer and van Hoesen (1959) suggested that the three large stars are Mars, Mercury, and Jupiter, and that the complex motif may represent a three-planet and lunar conjunction of July 7, 62 B.C. in the constellation of Leo (figure 243), which may also have been the date of Antiochus's coronation.

Figure 242. Tomb of Antiochus I

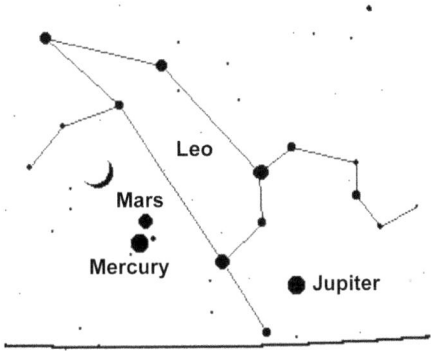

Figure 243. July 7, 62 B.C.

Complex Astronomical Designs on Ancient Coinage

The use of a zodiacal symbol similar to that on the tomb of Antiochus I is found on the reverse of several bronze coins of the Roman Emperor Julian II (360–363). A bull is depicted with a small star between its horns and a larger star above its shoulder (figure 244). In the spring of 360, Julian's troops rose in revolt against Constantius and proclaimed Julian II as Augustus. The depiction of the bull is fairly well understood: Julian II often slaughtered bulls as victory offerings to Mars, the Roman god of war. But what do the stars represent?

On May 4, 360, Venus joined Mars to form a single bright object (0.076° separation) between the horns of Taurus (the Bull) as the constellation set in the western sky. Two weeks

Figure 244. Julian II

earlier, Mars was between the horns and the brighter planet, Venus, rested on the bull's shoulder (figure 245). There can be little doubt that Julian saw this planetary conjunction as an omen for his victory and showed it on his coins.

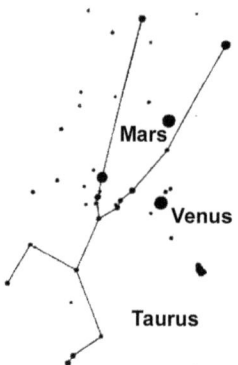

Figure 245. April 20, 360

Planetary conjunctions of this small separation are extremely rare and spectacular. For an adult, a fist extended at arm's length subtends about 10° of arc in the sky, and a thumbnail covers about 2°. The occultation (apparent fusion of two planets into a single object) requires that they be visibly separated by less than about 0.03°, unless one of the planets is exceedingly bright, as is Venus.

Constellations on ancient coinage were also depicted using stellar symbols. In 74 B.C. L. Lucretius Trio of the Roman Republic struck a denarius with the sun god Sol on the obverse and a crescent moon with seven stars on the reverse (figure 246). The seven stars cannot be the seven planets, because the moon is depicted as a crescent, but twice the year before, on October 11 and December 4, 75 B.C., the moon occulted the Pleiades star cluster. While the passage of the moon in front of the Pleiades is not rare, it is visually striking and happens infrequently enough to be noteworthy.

Figure 246. Lucretius Trio denarius

The presence of a crescent moon with stars does not always mean that the moon has joined a stellar group, but sometimes depicts a stunning pairing of the crescent moon with a close multi-planet conjunction. A rare bronze uncia of the Roman Republic, issued around 217–215 B.C., has a pellet within a crescent under two stars (figure 247). The pellet is the mark of the denomination of the coin, but the stars and crescent are probably the planets and the moon, respectively. Early in the morning of March 10, 217 B.C., a thin crescent moon rose next to a close conjunction of Mars and Mercury. Venus and Saturn were nearby, forming a beautiful sight (figure 248).

During 109–108 B.C. Man. Aquillius struck a Roman Republic denarius with the radiate

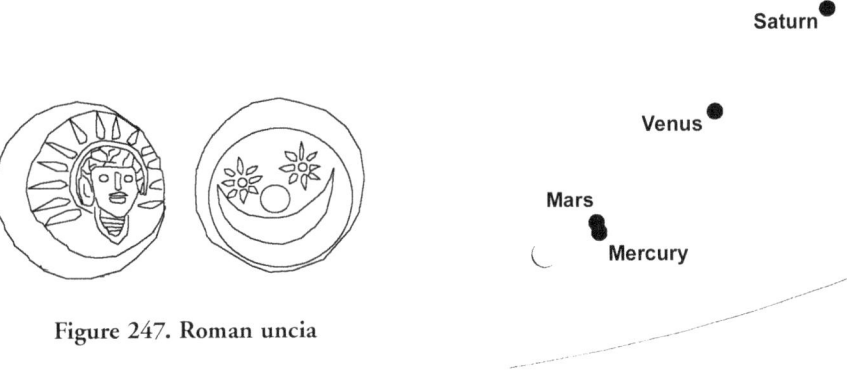

Figure 247. Roman uncia

Figure 248. March 10, 217 B.C.

head of Sol on the obverse and Luna in a biga (chariot with two horses) on the reverse (figure 249). Above the biga are a crescent moon and three stars and below it is a fourth star. Shortly after sunset on June 17, 109 B.C., a thin crescent moon was in conjunction with Jupiter, Mars, and Venus (figure 250). Mercury was just above the horizon, but may have been lost in the twilight. Saturn was close by in the southwestern sky. This would have been a spectacular sight, as the trio would have remained visible as the sky darkened into night. Either Mercury or Saturn could be represented by the star below the biga.

Figure 249. Man. Anquillius denarius

Figure 250. June 17, 109 B.C.

In 42 B.C. Clodius, a moneyer of the Roman Republic, issued a gold aureus and a silver denarius with the head of Sol on the obverse and a crescent beneath five stars on the reverse (figure 251). In the predawn southeastern sky on January 17, 44 B.C., a thin crescent moon rose beneath a rare conjunction of all five of the known planets (figure 252). Perhaps this extraordinary celestial event was recorded on this coin.

On a gold stater of Celtic Iceni struck between 45 and 40 B.C. the obverse of the coin contains opposing crescents with three pellets between the lunar cusps, and three pellets form triangles above and below the crescents. On the reverse, an annulet enclosing three pellets is depicted above and on a horse, with a star below the horse (figure 253).

The Iceni coin portrays quite a variety of astronomical symbols. Between 50 B.C. and 40 B.C. three of the known planets came together several times in various triangular configurations or close groupings, and the three pellets may represent any of these conjunctions. In 46 B.C. Venus, Jupiter, and Saturn formed a tight, 2°, triangular conjunction, and just after

86 Astronomical Symbols on Ancient and Medieval Coins

Figure 251. Clodius denarius

Figure 252. January 17, 44 B.C.

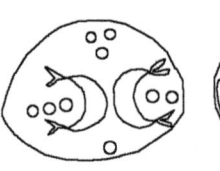 

Figure 253. Iceni gold stater

Figure 254. November 4, 46 B.C.

sunset on November 4 these three planets were visible in the southwestern sky next to a thin-crescent moon (figure 254). Could this be the design found on the stater?

Between 125 and 138 A.D. the Roman Emperor Hadrian issued denari with two, four, five, or seven stars above a crescent (figure 255). Other types contained a star within a crescent with a pellet below (figure 256). This latter type is also found without the pellet. The pellet does not denote the denomination of the coin on the denarius.

Figure 255. Hadrian

Figure 256. Star, crescent, and pellet

The designs could be based on several astronomical events. Lunar occultations of the Pleiades are certainly a possible explanation for the seven star and crescent design. For example, a thin crescent moon was near the Pleiades on October 29, 125, and again on March 15, 126, and a nearly full moon occulted the cluster on October 9, 127. But what are the explanations for the two-, four-, and five-star and crescent designs?

Conjunctions of planets with the moon may hold the answer. In September 131, all five of the known planets clustered in the constellation of Libra, and on September 11 and 12, a thin crescent moon was directly below the line of planets.

The two- and four-star and crescent types of Hadrian may also be explained by the same planetary conjunction. Mercury, Venus, Mars, and Jupiter formed the tightest grouping of the five planets, with all four coming together within a 5° circle on September 15. Between August 26 and October 28, 131, all four of these planets "danced" with each other. Each of the planets had a very close conjunction with each of the other three planets. The most spectacular of these two-planet conjunctions was between Venus and Jupiter, when on September 11, they were separated by only 0.3°. One day earlier these two planets had still been close together and directly above the thin crescent moon (figure 257).

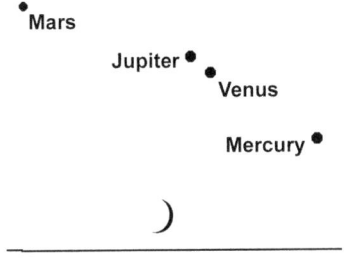

Figure 257. September 10, 131

The star, crescent, and pellet motif, first issued between 125 and 128, has an interesting celestial explanation. In spring 125, Venus surely would have drawn attention to the heavens with its conjunctions. In March 125 Venus set just after the sun along with Jupiter, and on March 8 the brilliant Venus and a lesser bright Jupiter were separated by only 0.26°. In April Venus was again involved in significant celestial groupings.

On April 21, 125, a solar eclipse was seen as a 26 percent partial eclipse in Rome, and may have darkened the sky enough to see Venus on one side of the solar crescent and Mercury on the other. This early morning eclipse passed directly over Roman-held Judea and Egypt, and Roman outposts throughout Asia Minor would have seen the two planets, one on each side of a thin solar crescent. The next day, on April 22, 125, an extremely thin, one-day-old, crescent moon set in the western sky next to Venus, with less bright Mars on the other side of the lunar crescent. Either of these two celestial views may be represented by a crescent with a star on one side and a pellet on the other.

The interpretation of complex astronomical designs can be subject to much debate. For example, Elagabalus, who was proclaimed as the Emperor of Rome by his troops in Emesa on May 16, 218, and was murdered in a praetorian camp on March 6, 222, struck a coin in Marcianopolis (now Devnya, Bulgaria) that has a thick crescent and three stars (figure 258). A number of scholars believe that the design represents the solar eclipse seen shortly after sunrise on October 7, 218. Mercury was next to the crescent of the partially eclipsed sun, and Saturn and Venus may have been visible some distance away (figure 259). However, a thick crescent would imply that the sky may not have been very dark there, and the visibility of all three planets is questionable. Alternatively, on the morning of January 26, 222, Venus, Mars, and Saturn rose in the east in a tight planetary conjunction with the crescent moon nearby (figure 260). Perhaps this is the event depicted on the coin.

Figure 258. Marcianopolis

88 Astronomical Symbols on Ancient and Medieval Coins

Figure 259. October 7, 218

Figure 260. January 26, 222

A provincial coin of Elagabalus struck in Sidon may depict the three planets involved in one of the above events. This coin has three military standards, or *vexilla*, on the reverse (figure 261). Each standard has an eagle resting on a crescent above two disks and a rectangular banner. Each banner has three pellets on it. Chapter 2 showed that in eastern regions, disks were often added to military standards to represent the planets as a pagan religious device.

Figure 261. Sidon

Complex Astronomical Designs on Medieval Coinage

The decline of artistic design in medieval coinage from that found on ancient coinage resulted in complex celestial events usually being represented by a combination of simple astronomical symbols added as additional marks to major design elements.

The Iberian Peninsula

Some of the most interesting complex designs are found on coinage of the Iberian peninsula. In 1096, upon the marriage of his daughter Theresa to Henry of Burgundy, King Alfonso VI of Leon-Castile gave Henry hereditary title to the province of Portugal as a vassal of the Castilian king. Henry died in 1112, and Theresa acted as regent for their son Alfonso Henriques.

Alfonso Henriques ruled from 1128 to 1185, and decisively defeated the Moors at Ourique around 1139. Portuguese legend states that a vision, a sign of the cross in the heavens, was taken as assurance of the victory. Perhaps the sign Alfonso Henriques saw was the conjunction of Mercury, Venus, Mars, and Jupiter in the evening sky in 1137. On June 20, the four planets formed a tight 7° grouping in the constellation Leo, and by July 2 had moved into a cross-like arrangement (figure 262). One month earlier, a 50 percent partial solar eclipse setting in the western sky was visible from the Iberian Peninsula, and perhaps increased Alfonso's awareness of celestial events.

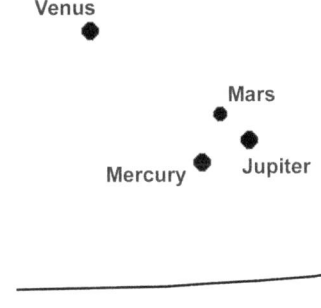

Figure 262. July 2, 1137

Based on his descent from Alfonso VI and his own conquests, Alfonso Henriques declared himself king of Portugal in 1143 and issued the first Portuguese coins. Two of his dineros were variations of a five-pointed star with a central pellet on the obverse and a long cross on the reverse. The star was to play a central theme as an immobilized design in subsequent medieval Portuguese issues. The 1079 total solar eclipse crossed southern Portugal and the annular eclipse of 1084 crossed northern Portugal. Either or both of these celestial events may have been depicted by the stellar motif and as omens for the recapture of Toledo in 1085 and middle regions of Portugal from Moorish control in 1094. The solar eclipses were of great significance, especially given the strong influence of Spanish Islamic astrology. Indeed, depiction of the eclipse on Christian coinage may have been used as propaganda to demoralize the Moorish population.

If the 1137 planetary conjunction were the basis of Alfonso's vision, then four planets may also be depicted on coinage. The coinage of Sancho I (1185–1211) contains dineros with a shield on the obverse and a cross on the reverse with either a pellet or alternate pellets and stars in each quadrant, perhaps a continuation of the eclipse star of his father's coins. The gold morabitinos of Sancho I, however, clearly depict five shields on the reverse, representative of the five Moorish kings who lost their lives at the great battle of Ourique. Each shield contains four pellets, and a star is in each of the four quadrants formed by the five shields. The obverse of the coin depicts the mounted king with raised sword and shield, and a star is found at the end of the legend (figure 263). Are the four stars or the four pellets on each shield representative of the four planets of the 1137 conjunction?

Figure 263. Morabitino of Sancho I

In Leon-Castle, Alfonso VII (1126–57) issued various coins with annulets, crescents, and stars (figures 264–265). The total solar eclipses of 1133 and 1140 and the annular eclipse of 1147 would have been seen as crescents on the Iberian peninsula. The magnificent annular eclipse of 1153 crossed the eastern part of Castile. Perhaps multiple eclipses are shown on the coinage.

Figure 264. Alfonso VII   Figure 265. Annulet type

Upon the death of Alfonso VII, Leon and Castile divided. In Leon, Ferdinand II (1157–88) issued a dinero with an annulet above a lion, probably in reference to the 1153 annular eclipse that had crossed part of Castile (figure 266). Ferdinand spent most of his reign trying to reunify the two kingdoms. Perhaps the design was a propaganda device to promote reunification under divine guidance. Three pellets are found beneath the lion. Two months prior to the 1153 eclipse, Mercury, Venus, and Mars were in a 3° triangular conjunction below the constellation of Leo.

Figure 266. Annulet type of Ferdinand II

Ferdinand II also issued a dinero with a mullet beneath two lions and a crescent and four pellets above (figure 267). On June 30, 1178, Mercury, Venus, Mars, and Saturn rose in the predawn sky in a 5° conjunction above the constellation of Leo, although Leo did not rise until after daylight and would not have been visible. On July 15 a thin crescent moon joined the four planets (figure 268). Two months later on September 13, 1178, the total solar eclipse that crossed the Spanish-French border was seen as an extremely thin crescent in Leon. The eclipse was situated just below the constellation of Leo.

Figure 267. Mullet, crescent, and four pellets

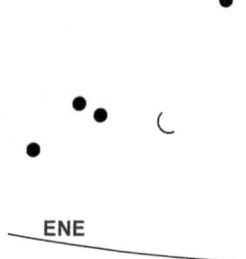

Figure 268. July 15, 1178

A dinero of Alfonso IX (1188–1230) has comet-like comb symbols in the quadrants of the obverse of the coin and a cross above the lion on the reverse (figure 269). In late spring of 1211 a comet was observed in the constellation of Leo.

Figure 269. Comet Type of Alfonso IX

Feudal France

Previous chapters have presented some examples of feudal French coinage depicting both solar eclipse and cometary symbols. Additional examples are presented here to demonstrate other symbol combinations.

Figure 270. Denier of Hugues III and obole of Hugues IV

Hugues III of Burgundy (1162–93) issued an interesting denier with annulets and three pellets in a straight line (figure 270). Subsequent oboles and deniers of Hugues IV maintained the design, but replaced the three pellets with three annulets. The annulets may represent the 1153 annular eclipse. In mid–March 1161 Venus, Mercury, and Mars came together in a straight line within a 2° separation (figure 271). This conjunction was probably taken as an omen for Hugues III and was represented by the three pellets. On November 3, 1162, Mercury, Saturn, and Venus came together in a triangle within a 4° circle. This second conjunction also may have played a role in the coinage design.

In the middle of the 12th century the bishop of Langres in Champagne issued a denier

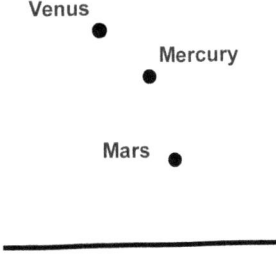

Figure 271. March 16, 1161

92   Astronomical Symbols on Ancient and Medieval Coins

with annulets, crescents, and a triangle (figure 272). These symbols may represent Halley's comet of 1145 or a comet in 1151 and the annular eclipse of 1147 or 1153. Both annular eclipses crossed Champagne. The comet of 1151 is significant in that John of Marmoutier, an abbey in the Loire Valley, refers to it following the death of Geoffrey, Count of Anjou: "Then having made bequests of grants, gifts, and charities, the death of so great a prince having been foretold by a comet, his body returned his spirit from earth to heaven" (Hallam 1995).

Figure 272. Denier of Langres

A denier issued in Orleans for Louis VII of France (1137–80) has an annulet in one quadrant and a pyramid in another quadrant of the reverse cross (figure 273). One of the mid–12th century comets and the annular eclipse of 1153 are probably represented. The four pellets within the city gates may depict Mercury, Venus, Mars, and Jupiter, which came together within 3° in June 1137, the first year of his reign, and Saturn replaced Venus in a 5° four-planet conjunction in April 1146.

Figure 273. Louis VII denier   Figure 274. Denier of Arnaud-Gausfred

A similar design is found on feudal coinage of Roussillon. Arnaud-Gausfred (1115–63) issued deniers and oboles with an annulet and pellet in one quadrant of the cross and an annuletted pyramid in the opposite quadrant (figure 274). The annular eclipse of 1153 passed directly over Roussillon, and during this spectacular event the sky probably became dark enough to see the brilliant planet Venus only a few degrees away (figure 275). The pyramid may represent one of the mid–12th century comets.

Now consider the deniers struck at Crepy by Philippe d'Alsace as Count of Valois (1156–83)

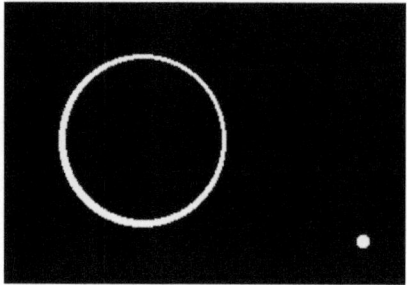

Figure 275. Roussillon January 26, 1153

and that of Matthieu d'Alsace struck between 1170 and 1174 (figures 276–277). In the first and third quadrants of the reverse cross, the design consists of a crescent and three pellets. The first question to be answered is the order of issue of these two coins, as the dates of Philippe's coins overlap those of Matthieu. Philippe's coin was probably issued first. Not only did he rule Valois from the time of his marriage to Isabelle, countess of Valois, in 1156, but any issue after 1178 may have contained mullets to represent the total solar eclipse of that year. Therefore Philippe's denier could have been struck as early as 1156 and Matthieu's denier could have been an immobilized design. The annular eclipse of 1153 was seen as a crescent in Crepy and covered about 80 percent of the solar disk. Venus was next to the crescent, Mercury was nearby, and Mars, about 30° away, is the likely choice for the third pellet.

Figure 276. Philippe d'Alsace

Figure 277. Matthieu d'Alsace

Germany

An interesting issue of King Henry II of Regensburg was issued between 1002 and 1014. Henry II was born in 973 and ruled as king from 1003 until his death in 1024. This issue replaces three pellets with a triangle in the first quadrant of the reverse cross and with an annulet in the third quadrant (figure 278). The year of his birth and the first year of his reign are almost the same as those of Stephen of Hungary on whose coins triangles were used to represent the comets of 975 or 995. Perhaps the triangle on Henry's coin does the same. The annulet might then represent the eclipse of 1010 seen in Germany as almost annular as it set in the western sky (figure 279).

Figure 278. Regensburg Penny

Figure 279. March 18, 1010

Unusual Designs

Aurora (northern lights) were sometimes recorded in contemporary historical records, but are difficult to portray on coinage. The same applies to lunar eclipses, because the moon is characterized by either a crescent shape in the partial phases or a reddish color during totality as the

94 Astronomical Symbols on Ancient and Medieval Coins

longer red wavelengths of light are refracted onto the moon by the Earth's atmosphere. A crescent lunar shape caused by a lunar eclipse cannot be distinguished on coins from the normal phases of the moon, and the portrayal of a reddish color was not possible. Coins did, however, have symbols that represented meteorites, supernovas, atmospheric solar circles, and glories.

Meteorites

Although not a common motif, ancients coinage sometimes portrayed meteorites, or "shooting stars," that land on the Earth (figure 280). Several of the Roman emperors of the second and third century depicted a sacred stone in a temple built for it.

The best known meteorite coin depicts the stone of Emesa. When Varius Avitus Bassianus was 14 years old, he was appointed as the high priest of the sun god Elagabalus, and as emperor he is known by that name. A great stone, or meteorite, was the symbol of the sun god cult, and the stone was located in Emesa, Syria. When Elagabalus moved to Rome as emperor in 219, he took the stone of Emesa with him, paraded it each year through the streets of Rome on a horse-drawn carriage, and built a great temple to house it. Some of his coins depict this annual event (figure 281).

Figure 280. Stone of Zeus Kasios   Figure 281. Stone of Emesa

Supernovas

Near the end of the life of a massive star, when its nuclear fuel is exhausted, the star will collapse internally, and a huge amount of energy will be released rapidly into space. This blast wave will appear to observers on Earth as a bright star-like object that will diminish in intensity after a number of days.

The great supernova of 1054 in Taurus (existing remnants are now referred to as the Crab Nebula) would most likely have been regarded as a significant portent in medieval Europe. This supernova was so bright that it was visible in daylight for 23 days. The naked eye observance of a supernova is so rare that the next two recorded supernovas were not until 1572 and 1604.

Although Chinese chronicles clearly record the 1054 "guest star as visible in the day like Venus, with pointed rays in all four directions," there are no known European records of the event. Modern historians suggest that religious prejudices forced medieval chroniclers to ignore the supernova, but it may have been recorded on medieval coinage.

Sear (1987) points out that on one type of gold coin of Constantine IX of Byzantium (1042–55), stars were added to each side of his bust (figure 282), and Sear attributed the stars to the 1054 supernova. Some have argued that the two stars represent a great schism in the church, but others argue that the schism was a series of actions rather than a single event, and that depiction of the supernova is more likely.

In Hungary, Andreas I (1046–61) modified the immobilized design of Stephen on one of his coins by changing the obverse to that of a central annulet and pellet with multiple rays extending from it (figure 283). This era in Hungarian history was one of acquisition of new territory and absorption of foreign colonists. The need to reaffirm the divine right of the king was a constant requirement. Clearly a spectacular event such as the 1054 supernova would have been required for a change in design, and would have been used to the king's benefit.

Figure 282. Constantine IX

Figure 283. Andreas I

In Anglo-Saxon England, a coin of Edward the Confessor bears a remarkable resemblance to the stellar motif of Andreas. The dating of late Anglo-Saxon and Norman English coins is subject to great debate and will be discussed further in chapter 7. Edward's expanding cross type is thought to have been issued between 1051 and 1054, but a reference to the 1054 supernova would place it after the celestial event. The reverse design features expanding rays emanating from a central annulet that usually contains a pellet (figure 284).

Figure 284. Expanding cross type

The expanding cross type was struck first on lightweight and then on heavy silver flans. On at least one of the early coins of this type, the central pellet has stellar-like rays (figure 285). Whether the rays are there by design or were caused by the production process is unclear.

Figure 285. Stellar-like pellet

Figure 286. Sovereign type

The sovereign type penny of Edward the Confessor, thought to have been struck between 1056 and 1059, has birds in the quadrants of the reverse (figure 286). The birds are described in numismatic literature as martlets or imperial eagles, but ravens are more likely because of their significant role in English mythology. For many centuries, ravens have guarded the Tower of

London, a symbol of the sovereignty and power of the monarchy in England. Legend predicts a great disaster for England if the ravens were ever to leave. The White Tower, built during the Norman period, is situated on a hill known as the White Mount, the legendary burial site for the head of Bran. As the legend goes, Bendigeid Vran ab Llyr of Wales, known as Bran, commanded that his head be buried there with his face toward France to prevent an invasion of England. *Bran* is the Welsh word for raven.

The ravens on the sovereign type penny could have an astrological association with the supernova. In European Mithraic cultures Corvus, the Raven, watched as Taurus died (Krupp 1991), or allegorically, as the supernova in Taurus faded during the next few years after the event. Curiously, the only known gold coin of Edward the Confessor is of the expanding cross type. The depiction of the 1054 supernova on coinage of Andreas of Hungary, Edward of England, and Constantine of Byzantium may be the only record in western cultures of this significant astronomical event.

Atmospheric Solar Circles and Glories

Common single and double rainbows, while often spectacular, are seen frequently enough that they were unlikely to have been considered as omens. However, unusual types of rainbows may have been seen and represented on coins.

One or more rings, or haloes (figure 287), can be seen in the daytime around the sun when sunlight is reflected off high-altitude ice. A similar phenomenon is more often seen against a dark nighttime sky when moonlight is reflected and one, or sometimes two, rings are observed. Solar haloes can be dramatic, often producing various colors and/or cross-like rays.

More often, only a portion of a solar halo is visible, or there are only irregular and bright solar reflections off atmospheric ice. Such reflections are known as sun-dogs, because they follow the sun as it moves across the sky. Sun-dogs may occur on one or more sides of the sun. Quite often, streaks of rainbow colors are also seen intermixed with the reflections. Crosses of light caused by sun-dogs have been reported. The observation of sun-dogs is not very common.

This author has seen less than a dozen in the past 30 years. The most spectacular sun-dog phenomenon seen by this author occurred in 1968, and a brief description of it may yield some insight into why such phenomena may have been considered as visions in ancient times. A blue sky appeared to be cloudless, although there must have been an very thin layer of high-altitude ice. The sun, along with two perfectly round solar disk reflections to its right, formed a large equilateral triangle in the sky, with one reflection above and the other below the sun. Between the two reflection disks was a brilliant and straight, vertical rainbow (figure 288). There actually appeared to be three suns in the sky!

Figure 287. Solar halo

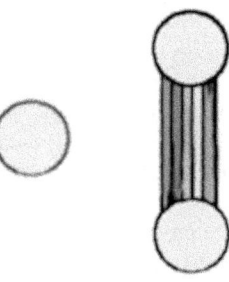

Figure 288. Sun-dogs

Four solar circles were recorded in the *Anglo-Saxon Chronicle* in 1104. This solar circle event may have been depicted on Norman coinage of Henry I. Three types of his pennies struck between 1106 and 1114 depict a diamond cluster of four annulets (figure 289). Other possibilities exist for the cluster and will be discussed in chapter 7.

Although the mechanism for the formation of a glory is quite different from that of a common rainbow, its visual appearance is similar to a rainbow in that all the same colors may be seen, but it is significantly different in that a shadow always appears in its center. Airplane pilots often see glories when the sun, the airplane, and fog or clouds are all in a straight line (figure 290). The shadow of the airplane is seen in the clouds, and an oval rainbow is seen around the shadow. Sometimes even the frozen exhaust of a jet can cast a shadow and form a long glory. Because this phenomena often is seen from aircraft, it is sometimes called a pilot's rainbow.

Figure 289. Facing bust type IX

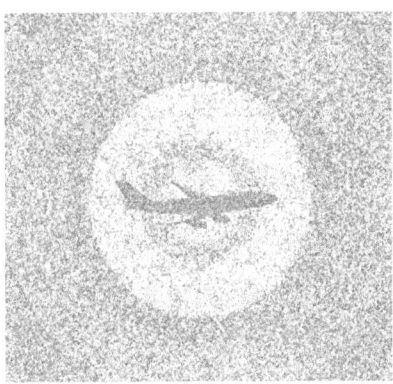

Figure 290. Glory

Glories can also be seen at ground level, with a low sun behind the observer that casts the shadow of an object, such as a statue, on fog or low clouds behind the statue. This type of glory can cause a rainbow to appear around the statue. As fog and clouds often may not reach all the way to the ground, the rainbow is more likely to be seen at the top of the statue, forming a colorful halo around its head. This type of glory is also known as a halo. As the statue blocks the central shadow from the observer, only the statue and halo are seen.

Solar haloes or glories may have been observed in ancient times behind the head of a person or statue and inspired a religious connotation. Perhaps the nimbus seen above the head of Christ and Christian saints on coinage stemmed from a solar phenomenon?

VISIONS EXPLAINED

Christogram

Fiery vision of the cross

Chronological List of Events

The following dates and events are relevant to the material presented in this chapter.

- 272   Constantine the Great (referred throughout as Constantine) was born.
- 303   Constantine's son Crispus was born.
- 306   Constantine was declared as Augustus by his troops.
- 310   Constantine saw a vision of the sun god Apollo, and chose Sol Invictus as his deity.
- 312   Constantine defeated Maxentius at Milvian Bridge and saw the first vision of the cross before the battle. The Roman senate presented Constantine with a statuette of Victory on a globe of world rule, which was preserved as the altar of Victory.
- 314   Constantine II was born.
- 315   The triumphal Victory Arch of Constantine was built in Rome.
- 317   Crispus, Constantine II, and Licinius II were appointed as Caesars. Constantius II was born.
- 323   Constans was born.
- 324   Constantine defeated Licinius and saw the second vision of the cross. Transformation of Byzantium into Constantinopolis, the new Christian capital of the Roman Empire, began. Constantius II achieved the rank of Caesar.
- 326   Crispus was executed.
- 328   The third vision of the cross was seen at the Danube Bridge.
- 330   Constantinopolis was dedicated.
- 333   Constans became Caesar.
- 335   Delmatius and Hanniballianus were elevated to Caesar and king of Armenia, respectively.
- 337   Constantine died. Delmatius and Hanniballianus were executed.

The Visions of Constantine

One of the great legends of all times has survived for almost 2,000 years, but until now, has never been explained. Although his troops were greatly outnumbered, in 312 Constantine sought to defeat Maxentius in battle as a major step in his quest to unify the Roman Empire. According to the legend, on the eve of the Battle of Milvian Bridge, Constantine and his troops saw a fiery vision of the cross in the sky along with the words *hoc signo victor eris* (under this sign you shall be victorious). Inspired by this vision, Constantine ordered the sign of the cross painted his troops' armor, and motivated by the promise of divine intervention, Constantine's army defeated the troops of Maxentius on October 28, 312. According to his chroniclers, the vision of the cross was the basis for Constantine's acceptance of Christianity: no longer were the Christians to be persecuted. The stage was set for the western world to find its religious majority.

Ancient writings recorded that similar visions appeared at least two more times during Constantine's reign. This is the story of the three visions and the answer to the question: what did Constantine and his troops really see?

Many excellent references are available that detail the history of the Roman Empire. The written record of official Roman chroniclers, the writings of pagan and Christian authors, and archaeological studies have all contributed to a detailed understanding of this period. Thus only a brief account is presented here.

At the end of the third century Diocletian was the Emperor. In 286 he divided the rule of the empire with Maximianus, who was elevated to the role of Augustus. Maximianus controlled the western half of the empire from Milan and Diocletian ruled the east from Nicomedia.

At this time, emperors considered that they derived their power from the gods and no longer from Roman citizens. Diocletian declared himself the earthly representative of Jupiter (figure 291), while Maximianus aligned himself with Hercules (figure 292).

Figure 291. Diocletian and Jupiter standing, holding a scepter and Victory with a wreath on a globe, and with an eagle at his feet

Figure 292. Maximianus and Hercules wearing a lion's skin headdress

In 293 Diocletian, created the Tetrarchy, a four-part division of military rule (figure 293). Two Caesars were picked to assist, and later to succeed Diocletian and Maximianus. Constantius Chlorus (figure 294), Constantine's father, was chosen in the west, and Galerius in the east. Under the seniority of Diocletian, Constantius was in charge of Britain and Gaul; Maximianus was responsible for Italy, Spain, and Africa; Galerius defended the middle and lower Danube; and Diocletian retained his responsibility for the east.

Figure 293. Leaders of the Tetrarchy performing a sacrifice in front of the camp gate

Figure 294. Constantius Chlorus

In 303 Diocletian issued three edicts against the Christian church, reversing the position toward the church that had followed the edict of toleration Gallienus had issued in 260. Church property was confiscated and destroyed, clergy were arrested, and Christians were declared outlaws and were forbidden to congregate. Of the four rulers, only Constantius did little to enforce the edicts. The other three treated the Christians harshly.

Two years later Diocletian retired because of ill health, and forced Maximianus to step down as well. Constantius and Galerius moved to their new positions as Augusti, with Constantius as senior. Under the direction of Diocletian, the two appointed two new Caesars, Severus and Maximinus Daia, to reconstitute the Tetrarchy, but Constantius was ordered to pass over his son, Constantine, and Galerius did not choose his son-in-law, Maxentius, who was the son of Maximianus. Resentment prevailed, but Diocletian's moral authority continued to hold the empire together.

In 306 Constantius died at York after being victorious in battle in northern England, and his troops declared Constantine as Augustus. With the retirement of Maximianus under protest, an explosive political situation had been created. To ease the tension, Galerius recognized Constantine as Caesar and promoted Severus to the rank of Augustus; however, political unrest among the rulers continued.

Maxentius revolted in Rome, invited his father to re-assume his title, and Maximianus eagerly accepted. Galerius ordered Severus to regain control, but Severus was captured and put to death in 307. The empire was on the brink of civil war.

A conference was called at Carnuntum in 308, and Licinius, a general of Galerius, was appointed emperor in the west. Maximianus was ordered to step down for the second time, and Maxentius was declared a public enemy. Constantine and Maximinus were angered by the promotion of Licinius. Over the next few years a series of alliances and attempts to maintain the Tetrarchy took place. Civil war had been averted, but the situation was still tenuous.

When Galerius died in 311, Licinius assumed control of Asia Minor and the lower Danube. Constantine still held Britain and Gaul, and Maximinus controlled Syria and Egypt. Maxentius remained in control of Italy and other western provinces. The stage was now set for battle.

Constantine had already begun his move to control the empire. A year earlier, Constantine had announced fictitiously that his father was the illegitimate son of Claudius Gothicus, the Roman emperor who had died in 270, thereby proclaiming an hereditary right to the entire empire.

In spring 310 Maximianus rebelled against Constantine and declared himself emperor for the third time. Constantine besieged him at Massilia, and Maximianus was forced by the shame of defeat to commit suicide. On the journey to Massilia near a sanctuary of the sun god, Apollo, Constantine received news of the collapse of a barbarian uprising on the Rhine. According to a chronicler, Apollo appeared before Constantine along with Victory and offered him four laurel wreath crowns, with each being a symbol for 30 years of success. Thus the first recorded vision of Constantine had a pagan reference.

With the death of Maximianus, Hercules lost his status as a political necessity, and his image ceased to be struck on Constantinian coinage. Constantine declared the sun god Sol Invictus as his deity, turning from Jupiter as well (figure 295). The gods who had supported the Tetrarchy were

Figure 295. Constantine and Sol (310–313)

no longer needed to support the political aspirations of the future ruler of the entire Roman world.

After Galerius died, Constantine formed an alliance with Licinius to move against Maxentius, and then began his military campaign to increase his control with the annexation of Spain. Italy was next. With an army only one-fourth the size of that of Maxentius, Constantine marched south through northern Italy. On October 28, 312, the two met in battle at Saxa Rubra, nine miles north of Rome.

Inspired by the heavenly vision of the cross, Constantine and his troops were victorious. As Maxentius retreated across the Milvian Bridge, he drowned in the Tiber River. With this victory, Constantine claimed the western Roman Empire. He was convinced that his victory at Milvian Bridge had been the result of divine intervention. Two different accounts relate the legend of this heavenly vision.

According to his biographer Eusebius, who wrote the *Life of Constantine* after the emperor's death, Constantine had seen the vision of a fiery cross in the afternoon sky and the inscription *hoc signo victor eris*. In his sleep later that night, Christ appeared and commanded him to make a likeness of the cross and use it as a safeguard in battle. Eusebius described the cross-like sign, often referred to as a *Christogram*, as the Greek letter *Rho* superimposed on the letter *Chi* (figure 296).

In contrast, Lactantius, tutor to Crispus, Contantine's eldest son, wrote that the vision occurred at night when Constantine was encamped near Saxa Rubra. As in Eusebius's version, Constantine was directed to place the sign of God on his shields and ordered the Christogram to be placed on his soldier's helmets and shields (figure 297). Lactantius wrote his account only a few years after the battle, and his description of the *Chi-Rho* symbol varies slightly from that of Eusebius. According to Lactantius, the symbol was a rotated *Chi* with a loop at the top (figure 298).

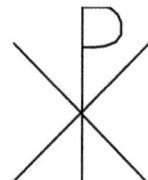

Figure 296. Chi-Rho of Eusebius

Figure 297. Chi-Rho symbol on helmet of Constantine (315)

Figure 298. Chi-Rho of Lactantius

The exact appearance of this remarkable vision is less important than the influence it had on Constantine and his troops. The *Chi-Rho* became the predominant religious and political symbol of the vision that initially influenced the toleration, and later the conversion, of the Roman Empire to Christianity. I will show that each form of the Christogram may have been observed in separate visions.

The *Chi-Rho* symbol was not new, nor was it new to coinage. On a coin of Ptolemy III (246–221 B.C.) of Egypt, a *Chi-Rho* was included below the eagle (figure 299), either as an abbreviation, or as some scholars suggest, as a combination of two ancient sun symbols.

Figure 299. Chi-Rho on coin of Ptolemy

Early in 313, Constantine and Licinius met in Milan and issued an edict that granted complete religious freedom to the empire's subjects. While the two were meeting in Milan, Maximinus crossed from Asia into territory held by Licinius, where Licinius defeated him on April 30, 313, and took full control of the eastern Roman Empire. For the next 11 years, Constantine and Licinius shared a joint rule that was strained to the extent that they sometimes found themselves on the battlefield facing each other.

In 317, in an effort to ensure their succession, Constantine's two oldest sons, Crispus and Constantine II, and Licinius's son, Licinius II, were appointed as Caesars, but religious and political differences continued. Even though Constantine continued to support pagan belief in the sun god, he adopted as his standard, the Labarum with a gold wreath encircling the Christogram. Some believe that Constantine considered the campaign against Licinius as a religious war to base the unity of the empire on Christianity.

After more than a decade of strained relations, Constantine marched into Thrace, and on July 3, 324, was victorious at the Battle of Adrianopolis. Licinius was forced to retreat 130 miles back to Byzantium. Under the command of Crispus, Constantine's fleet won naval victories and blockaded the coast of Byzantium, forcing Licinius to flee from the city.

The reign of Licinius ended at Chrysopolis on September 18, 324, when he was captured by Constantine's army and then executed the following year. After Chrysopolis, Constantine marched virtually without resistance toward Byzantium, and the city quickly surrendered. Work soon began to transform the city into the new capital of the empire, Constantinopolis. Five years later, on May 11, 330, the new capital was dedicated.

Soon after the defeat of Licinius, another son of Constantine, Constantius II, was named as a Caesar, and the same rank was bestowed on Constantine's youngest son Constans in 333. But not all was joyous in the life of Constantine. In 326 his second wife, Fausta, jealous of her stepson, Crispus, fabricated evidence of treason that led to his execution. Upon learning of the treachery, Constantine had her thrown to her death in boiling water.

In 335 two of Constantine's nephews, Delmatius and Hanniballianus, were elevated to the role of Caesar and king of Armenia, respectively, and the empire was divided among the five cousins with supreme power in the hands of Constantine the Great, who was the sole Augustus until his death on May 22, 337. A comet was said to have predicted his death, and coinage that commemorates his passage to heaven portrays a hand extending from heaven (figure 300). The two nephews were then put to death, and Constantine's three remaining sons became joint rulers of the Roman Empire.

In 336, during a lengthy oration, Eusebius referred to the sign of the cross as a harbinger of victory, a divine vision of the Savior that had often shone on Constantine. According to an anonymous chronicler, after his mother, Helena, had returned from a pilgrimage to Jerusalem, Constantine erected three great crosses to represent each of the three times one had appeared to

Figure 300. Constantine in quadriga with the hand of God above

him as a vision in time of war. The first was prior to the Battle of Milvian Bridge; the second was during his conquest of Byzantium; and the third was in 328, when he built his famous bridge across the Danube River in Scythia.

The second vision was recorded as occurring while Constantine marched from Nicomedia and fought against the Byzantines after slaying Licinius. However, most historical records show that the Byzantine population was not a threat after Licinius was defeated at the Battle of Chrysopolis. This second vision was probably manifested at the Battle of Adrianopolis.

Thus, the journey of Constantine the Great had ended, but the legend of his visions has endured, and now they can be explained. Constantine did see the Christogram in the heavens above.

Visions Explained

As discussed, Constantine had three heavenly visions of the cross. In this chapter, first, second, and third refer to visions of the cross and not to his earlier vision of Apollo.

Just as the Roman emperors before him had done and as those who followed him would do, Constantine used astronomical symbols as propaganda devices. Thus coins struck during the reign of Constantine will be one of the sets of evidence used to explain his visions. In addition to the written historical record and the analysis of coinage design, the remaining evidence is the appearance of the heavens when the visions occurred.

For the month prior to Constantine's victory at Milvian Bridge, Mars, Saturn, and Jupiter formed various shapes in a three-planet conjunction in the southwestern sky just after sunset. Venus was visible just above the horizon. At their tightest grouping, on October 9, 312, the trio of planets were in a 3° triangle. On the eve of the battle all four planets formed a nearly straight line. These planets would have commanded the attention of Constantine and his troops as they marched toward Rome and made camp to prepare for the upcoming battle. When the October 27 night-time sky is recreated showing the brighter stars as well, the first vision is revealed (figure 301).

On the eve of the battle, Venus was in the constellation Sagittarius and Mars was in Capricorn, with Saturn and Jupiter in between. The line of planets and the bright stars of Capricorn formed the *Chi*, and one arm of the cross extended into the constellation of Aquila, where the bright star Altair started the loop at the top to form the *Rho*. While one can argue how to "connect the dots," the bright stars and planets offer a very likely choice. DiMaio, Zeuge, and Zotov (1988) depicted the loop of the *Rho* to be much larger involving additional stars to those shown in figure 301. Nevertheless, in both constructions, the *Chi-Rho* Lactantius described was clearly visible.

Just after the sun set the night before the battle, the four planets would have been visible

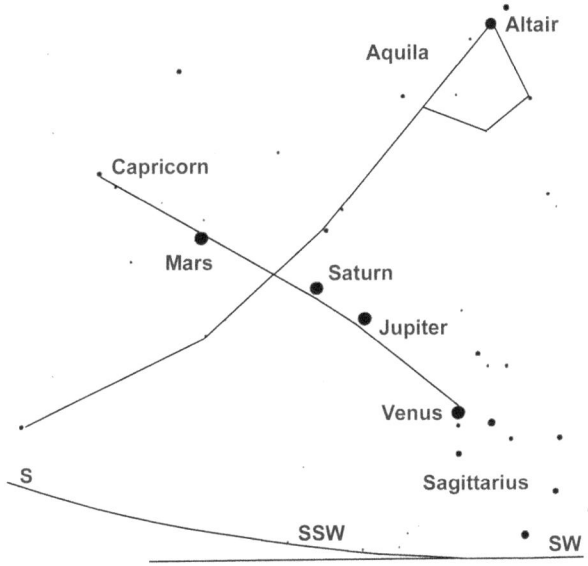

Figure 301. First vision of the cross, October 27, 312

in the south-southwestern sky, with Mars almost due south. The troops of Maxentius were encamped nine miles north of Rome at Saxa Rubra, directly south of Constantine along the Flaminian Way and with their backs to Milvian Bridge, which was several miles away. The next day, Constantine led his troops to the south in the direction where they had seen Mars, the Roman god of war, at sunset the evening before, to attack Maxentius's western flank. Victory lay ahead of them, *under the sign of the cross.*

Even after this first vision of the cross, Constantine was still strongly committed to his new deity, Sol Invictus (figure 302). If he had interpreted this vision as a Christian sign rather than as a pagan sign, he certainly did not fully embrace Christianity as a result of it. Constantine continued to demonstrate his belief in the sun god for at least another decade. Indeed, his triumphal arch, built in 315, was dedicated to Sol and not to Christ.

Figure 302. Constantine and Sol (313)

Furthermore, Constantine did not consider the Christian cross as a religious symbol until after Helena had returned from the Holy Land with a relic of the cross. Prior to that, a cross was merely a tool of execution to Constantine. More likely, Constantine saw a combined *Chi* and *Rho* in the heavens and viewed it as a blessing of both Christian and pagan gods. Eusebius, who wrote that the vision was that of the Christian cross, used the vision for his own political purposes to spread the word of Christianity.

However, Constantine was as skillful a politician as he was a military commander. He did what he had to do to gain the support of his troops, the populace, and whomever he needed for an ally at the time. In 307 Constantine had divorced Minervina and married Fausta, the daughter of Maximianus, as a political move, and he had made an alliance with Licinius against Maxentius when he had needed to have a military ally. Thus Constantine might just as pragmatically have accepted both pagan and Christian beliefs to gain the support he required.

As early as 315, coinage of Constantine depicts the Christogram as described by Eusebius (figure 297); however, as late as 336, some Constantinian coins show the *Chi-Rho* as described by Lactantius (figure 303). Thus the portrayal of the exact *Chi-Rho* of the vision was not of primary importance. Constantine had no reason to have the symbol depict exactly what he saw, as the Christogram, *any Christogram*, was sufficient for him to gain the support of the growing Christian population. The reference by Eusebius of the vision of the fiery cross in the sky as a daytime event was probably propaganda. Perhaps it was a way to combine a cross with the sun and thereby provide a transition from Sol to Christianity.

Figure 303. Chi-Rho of Lactantius

Furthermore, in the *Ecclesiastical History of Philostorgius*, the monk John wrote an account of the martyrdom of Artemius, in which Artemius appeared before Julian II and proclaimed that he had been present during the battle against Maxentius and had seen the sign of the cross in the daytime and the writing of the victory in the stars. Thus Artemius re-stated the daytime vision, and confirmed the sighting of the night-time event not as a dream, but as a celestial event.

These accounts do not provide enough evidence to build a circumstantial case for what was seen. Eusebius and Artemius had religiously political gains to achieve. The account written by Lactantius was probably closer to reality. Constantine and his army did see what appeared to be a *Chi-Rho* in the evening sky, and at least some of the troops wore a Christogram on their armor when they went into battle against Maxentius, but on October 28, 312, the symbol may have been given a pagan meaning, a sign from Mars, the Roman god of war.

Note that Constantine issued another coin similar to his 310–313 Sol Invictus piece, but this time Mars was the deity honored (figure 304). A star is located on the helmet of Mars. The star may be a reference to the sun or Sol, or it may mean that a star was also used as a symbol for Mars.

Figure 304. Mars   Figure 305. Stars and crescent

A rare variety of the bust of Mars on this coin may help to explain the significance of the star on his helmet (figure 305). On this variety not only is a star found on the ear flap of the helmet, but five additional stars and a crescent are displayed on the bowl of the helmet. A wavy line separates the ear flap star from the other symbols.

This unusual design has at least two possible explanations. It may simply be a representation of the five visible planets, the moon, and the sun. In this case, the star on the ear flap would most likely represent the sun, as it is separated from the other stars, and it would be more plausible to group the five planets together than to group the sun with four of the planets, but another possibility exists. The design may represent an actual celestial event where a crescent moon joined a conjunction of the five planets.

This type of event happened three times during 310–313 when the coin was issued. The first time was in September 310, and the third time was in December 312, but the conjunction of November 18, 312, is the most likely event to be depicted (figure 306). Just three weeks after the Battle of Milvian Bridge, Venus had moved southward to form a tight triangular conjunction with Jupiter and Saturn, with a thin crescent moon directly below. These three planets and the moon are in the same configuration as the central part of the helmet design, and Mars and Mercury teamed with Altair and the other three planets to form a spectacular *Chi-Rho* in the heavens just after the sun had set below the horizon. Could the wavy line represent the horizon?

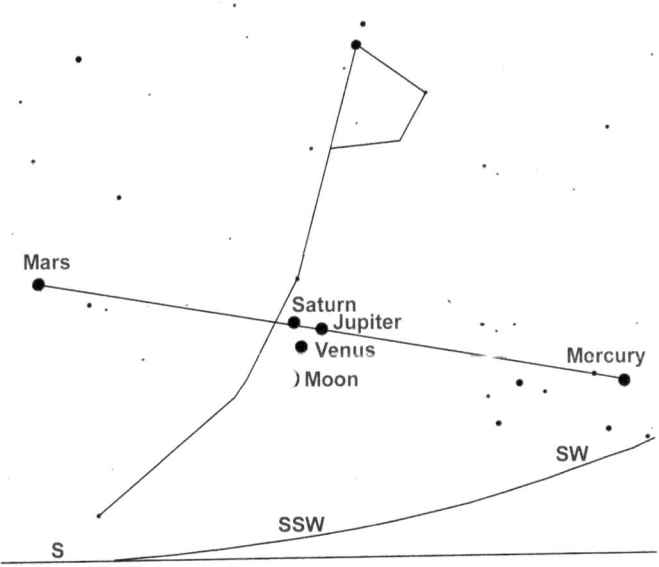

Figure 306. November 18, 312

One of these two possibilities must be correct. Either the design is a generic representation of the five planets, the sun, and the moon or it is a representation of an actual celestial event. In either case, this coin demonstrates that actual planets were depicted on at least some of the Constantinian coins, and that the star symbol could be a representation of a celestial body in addition to being a deity. The rarity of this helmet design variety suggests that it may have been struck for only a short time, which could place its date of issuance in 313, and thus consistent with the November 312 celestial event.

Propaganda on Constantinian coinage clearly refers to both pagan gods as well as the

Christogram, although coins with a *Chi-Rho* and stars on the helmet of Constantine (figure 307) are much rarer than similar pieces with only stars (figure 308). The Christogram pieces confirm that the *Chi-Rho* vision was used as a political device, whether or not it was actually observed. The two stars on Constantine's helmet probably refer to Sol, but references to both Sol and Mars or to the Dioscuri (Castor and Pollux) are also a possibility.

Figure 307. Chi-Rho and stars

Figure 308. Two stars (319–320)

Just as a jigsaw puzzle needs to have all the pieces in place before the entire picture is revealed, analysis of the visions of Constantine requires this discussion to continue. The second vision of the cross occurred in 324, when Constantine sought to defeat Licinius and once again, place the entire Roman Empire under a single command.

The date of Constantine's second vision was not clearly recorded in the historical record. Pagan writings placed it during a battle with the Byzantines after the death of Licinius, but most scholars believe this to be incorrect. Constantine and Licinius fought two major battles in 324, and re-creating the heavens during each of them is a simple task. The second vision of the cross is immediately apparent. Just before dawn on the morning of the Battle of Adrianopolis on July 3, 324, Constantine and his army were treated to another spectacular heavenly sight (figure 309). Mercury, Saturn, Mars, and Venus were in a straight line, the first three in Gemini and Venus in Taurus next to a thin crescent moon. Castor in Gemini and Betelgeuse in Orion formed the other line of the *Chi*, and the bright stars of Auriga served as the loop of the *Rho*. This time, the Christogram that Eusebius has described after the death of Constantine, the symbol seen throughout the Roman Empire, was clearly visible in the heavens above.

The third vision of the cross occurred when Constantine built his famous bridge across the Danube. The date of this event has not been found in the written record, but most scholars believe that it was in 328. Under the assumption that the *Chi-Rho* would involve yet another three- or four-planet conjunction, computer analyses were used to search for such an event in that year. Although a few possible planetary conjunctions took place that year, only one formed a *Chi-Rho*. Remarkably, the constellations that served up the stars for the third vision are the same as for the second one. Shortly after twilight on the evening of April 20, 328, Constantine and his troops would have seen the heavens as shown in figure 310.

Clearly all three visions, as documented in the written record, can be reconstructed as celestial events, and were portrayed on Constantinian coinage as the Christogram. However, the final piece of the puzzle still needs to be put into place. If the *Chi-Rho* sightings were also taken as pagan omens, then they should also have been recorded on coinage using astronomical symbols.

Aside from use of star symbols and the rare use of the Christogram, coinage of Constantine offered little evidence of the first vision until 319. Seven years after the victory over Maxentius, several mints issued coins with a seated captive on each side of a standard bearing the letters *VOT XX*. To the left of the standard was the Christogram (figure 311). Then in 321, Constantine began to reinforce his political claim to the entire empire.

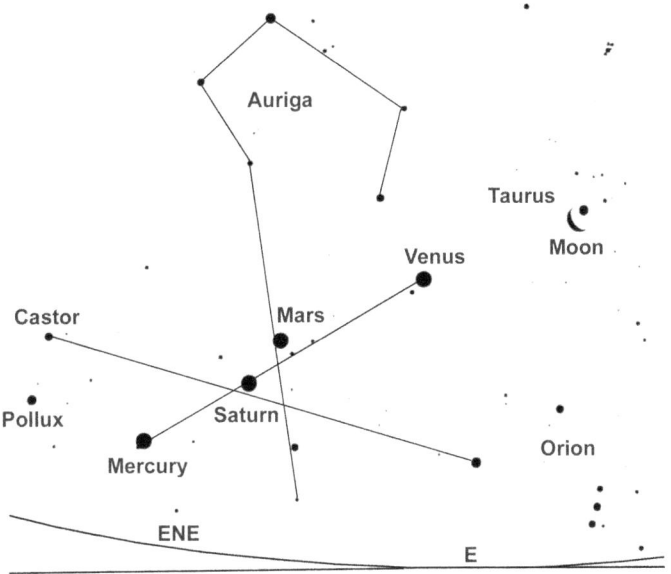

Figure 309. Second vision of the cross (July 3, 324)

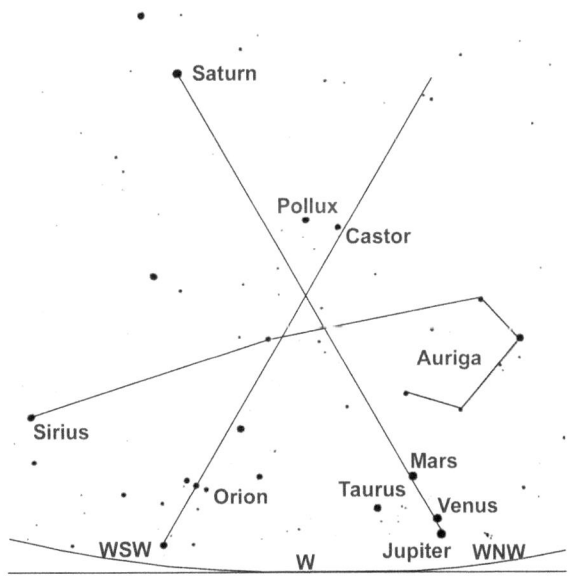

Figure 310. Third vision of the cross (April 20, 328)

From 321 to 324, numerous mints issued bronze coins with the victory altar the Roman senate had presented to Constantine in 312 (figure 312). Above the globe of world rule are three stars. The reverse legend, *BEATA TRANQVILLITAS*, combined with the military bust on the obverse of the coin, suggests an expectation of peace and tranquility as a result of military conquest.

The globe and three stars are of great interest to this discussion. Although the design on the globe has several variations, the usual motif is a single- or double-line cross with decorative

Figure 311. Captives type

Figure 312. Altar type

Figure 313. Typical globe designs

pellets (figure 313). Most likely, the lines represent the Zodiac and the celestial equator. The crossing point is an equinox. The globe with celestial arcs was a common motif of the ancient religious cult of Mithraism, which began its spread throughout the Roman Empire in the first century A.D. and reached its peak in the third century. Adherents of Mithraism continued to practice it until the end of the fourth century, and Constantine may have used this globe to rally the support of followers of Mithraism, especially as they made up a large part of the army.

Die engravers at the various mints decorated the globe on the coin at their discretion. In London, one engraver placed three pellets on each side of the bottom half of the globe (figures 314–315). Perhaps this was purely decorative, or maybe the three pellets carried the same meaning as the three stars. Or maybe this was a reference to the three-pellet motif used on Celtic coinage 300 years earlier.

Figure 314. London Mint (321–322)

Figure 315. Globe detail

Could the three stars above the altar be representative of Mars, Saturn, and Jupiter? These three planets were in conjunction for the entire month prior to the Battle of Milvian Bridge. Some astrological interpretations suggest that four planets in alignment may have been taken as a bad omen, and Constantine may have wished not to refer to the presence of Venus as part of the Christogram to reduce the number of planets seen in the vision. There would have been some justification for the exclusion of Venus, as it was situated in a different constellation and was not a part of the tight grouping of the other three planets. Indeed, some have suggested that the *Chi-Rho* itself was a mechanism used to diffuse the planetary connotations.

A better explanation of the three stars is that they represented Sol, or Mars, and the Dioscuri. In 321 the orator Nazarius stated that during Constantine's victory over Maxentius, two young men on horseback performed such great feats of gallantry that Constantine ordered them to be found and rewarded. The two young men were identified by Constantine as the Dioscuri.

The *BEATA TRANQVILLITAS* coin was issued prior to the second vision of the cross. In September 320, Mars formed an 8° conjunction with Castor and Pollux. The trio rose in the east shortly before midnight and moved across the night-time sky. After moving eastward away from the twin stars of Gemini, Mars reversed its path to make a closer revisit to the Dioscuri. On January 4, 321, Mars, Castor, and Pollux rose in the east in a straight line as the sun set, and were visible all night long (figure 316). By the end of the month they were in a 5° triangle (figure 317). The celestial combination of the Roman god of war with the mythical brothers who helped Constantine defeat Maxentius may have signaled the time to prepare for battle with Licinius.

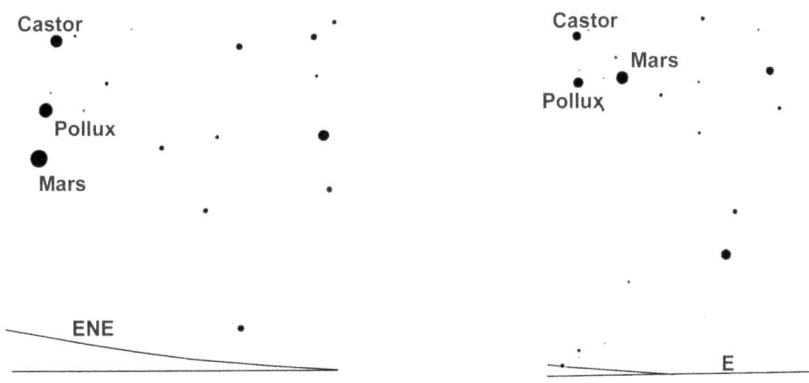

Figure 316. January 4, 321   Figure 317. January 30, 321

Just how rare were these conjunctions of Mars and the Dioscuri (Gemini)? Mars and Gemini set together in the west shortly after sunset in the spring of 313, 315, 317, and 319. On April 30, 317, Mercury, Venus, Jupiter, the crescent moon, Mars, and Gemini were visible in the western sky for a few hours after sunset. But Mars and the Dioscuri had not risen together at sunset in the east since 289. Mars came close to Gemini as a rising sign in the fall of 318, but backed away. Was the rising of Mars and the Dioscuri in the east at sunset important to Constantine?

From an astrological perspective, planets rising (ascending) in the east were in an especially powerful position. In January 306 the Dioscuri and Jupiter rose together in the east at sunset (figure 318). Six months later, Constantine became Caesar. In 310 during a speech that proclaimed a link between Constantine and Claudius, the orator emphasized that the gods favored this hereditary succession.

Another Constantinian coin of the period suggests that two of the three stars above the altar may well be the Dioscuri. A rare silver, medallic, miliarense piece struck around 320–321 has an altar flanked by two stars (figure 319), suggesting a reference to the Dioscuri. This coin was probably struck as a commemorative piece on July 25, 320, six months to a year prior to the three-star type and before the Mars and Gemini conjunction.

The meaning of the globe and three stars may be debatable, but their significance is unquestionable. Constantine was proclaiming his divine right to rule the world, and he was making his appeal to all his subjects: pagan, Christian, and Mithraic. He was setting the stage to gain the support he needed to defeat Licinius.

After Licinius's defeat in 324, a single star was added

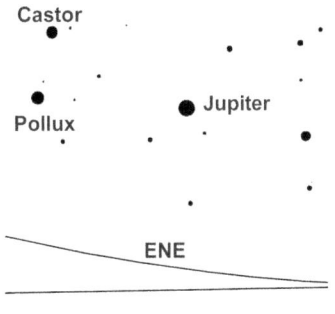

Figure 318. January 4, 306

112   Astronomical Symbols on Ancient and Medieval Coins

Figure 319. Two stars and altar type

Figure 320. Camp gate

at the top of the common camp gate coinage design (figure 320). Whether it represented Sol or Mars, the message was clear: Constantine was honoring his pagan gods. The star and camp gate design was issued at various mints until 329.

The first so-called Christian coinage of Constantine was struck in 327 at the new mint in Constantinopolis (figure 321). The reverse of a bronze coin portrays the Labarum, or Christian standard, piercing a serpent (evil, or perhaps Licinius). The inscription, *SPES PVBLICA*, means "the hope of the people." Above the Labarum is the Christogram. On the Labarum is only one symbol: three pellets.

Figure 321. Labarum

The popular explanation for the three pellets is that they represent Constantine's sons; however, this interpretation is probably incorrect. More likely they represent a three-planet conjunction. Consider the following. The Eusebius version of the first vision of the cross, written a decade after this coin had been issued, contained a detailed description of the Labarum. According to Eusebius, the Labarum was made the morning after the vision as a protective symbol to be carried into battle. His description of the first Labarum is questionable. He described it as being a long spear with a crossbar forming a cross that was covered with gold. A square banner embroidered with gold and jewels hung from the crossbar. Above the banner and crossbar was a portrait of Constantine and his children. At the top was a golden wreath with the *Chi-Rho* in the center.

The translation of Eusebius's *Vita Constantini* vary, and many versions refer to portraits (or medallions) of Constantine and his sons, while a few refer to Constantine and his two eldest sons. A possibility is that the original text of Eusebius and/or translations were biased by assumptions of what was thought to be depicted on the Labarum coin of 327. The translation that refers to his two eldest sons was probably contrived to fit the coin design.

Some scholars even argue that other historians finished *Vita Constantini* after the death of Eusebius. Assumptions and erroneous information may also exist in the original text. This discussion, however, can continue on with the assumption that the *Vita Constantini* does contain the description of a Labarum that existed during Constantine's reign.

Likely a crude banner was made on the morning of the October 28, 312, but certainly there was no time to prepare an elegant one fashioned of gold and precious jewels. And in 312 Constantine had only one son, Crispus, so his children could not have been portrayed, nor could his sons, nor could his two eldest sons. As Eusebius first saw the Labarum years later, the description of the standard that went into battle against Maxentius must have been embellished to reflect a Labarum that was fashioned long after the Battle of Milvian Bridge.

Even so, could this coin still represent the Labarum described by Eusebius? The Labarum on the coin of 327 only had three pellets. Constantine had three living sons by then. If the pellets represented the portrait of Constantine and his children, or his sons, four pellets would have been used, or perhaps even five, if Crispus, who had been executed in 326, were included. Also the pellets are on the banner, not above. Some translations, however, place the medallions as hanging below the banner, so the location of the portraits is probably not as important as the number of sons.

Additional evidence suggests that the pellets represent planets. In one of his volumes on Roman coinage, Cohen (1880) describes this coin as having either three pellets or three stars. Although verification of the existence of this coin with three stars remains to be shown, stars would rule out the pellets representing portraits. Perhaps Cohen could not determine if pellet features on worn copies of this coin may have been stars, but there is ample justification (the altar type) exists to believe that three stars were a significant design motif and therefore may have been used on this coin.

Furthermore, an example of this *SPES PVBLICA* coin found in the Fitzwilliam Museum in Cambridge, U.K. has three annulets instead of pellets. Although it is most likely a forgery, it may have been struck contemporaneously using dies copied from an actual coin. This is significant in that it demonstrates that various symbols were used on the banner of this coin type, therefore lending more credence to the possible stars type.

Some argue that the three pellets on the Labarum coin represent Constantine and his two eldest sons that were Caesars. After the death of Crispus and the defeat of Licinius, only Constantine II and Constantius II remained at that rank. This would not support the Labarum as described by Eusebius, but might be a possible explanation for the coin. This argument that the pellets only represent Caesars, however, would be inconsistent with another coin Constantine issued six years later (figure 322). This gold coin shows him holding a large standard bearing five annulets, along with two smaller standards, each bearing a single annulet. This coin was issued in 333 and does not involve a *Chi-Rho* symbol. Perhaps the five annulets represent Constantine and his sons, but this is doubtful, as Crispus had been dead for seven years and no other posthumous coin design references to Crispus exist. It also could not represent the four Caesars and the king of Armenia, as Delmatius and Hanniballianus were not elevated to their ranks until 335. Perhaps a celestial event, a divine omen, was represented on the standard. In the early morning of October 5, 332, a thin crescent moon and all five of the known planets had risen in Virgo, near the bright star Spica, in a 5° conjunction (figure 323).

Chapter 2 discussed the representation of planets on military standards by circles and disks, especially in the eastern regions of the Roman Empire. Chapter 5 presented a coin struck by Elagabalus that has military standards with banners bearing three pellets that may refer to one of two three-planet conjunctions. Earlier this chapter refers to a Greek coin of Egypt, struck more than 500 years prior to the Labarum coin of Constantine which depicts a *Chi-Rho* symbol.

Thus the assumption that the religious *vexillum* of Constantine, the Labarum, may have been constructed in a similar fashion would not be unreasonable. The form of the Labarum was almost exactly the same as that of eastern religious *vexilla* of the third century. These standards had a religious symbol at the top and a heavy banner below that was inset with precious stones

114   Astronomical Symbols on Ancient and Medieval Coins

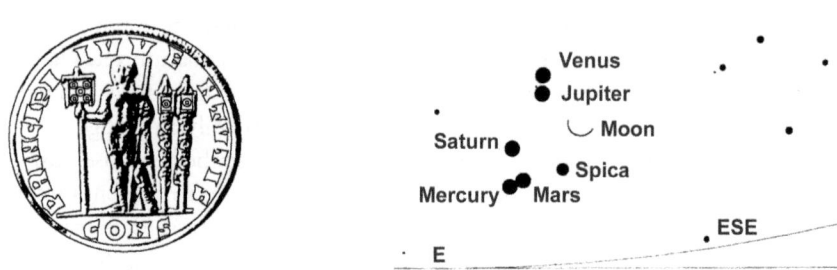

Figure 322. Constantine and standard

Figure 323. October 5, 332

that symbolized the heavens. Thus the three pellets on the Labarum coin and the five annulets on the coin of 333 would represent planets.

The Labarum coin with the three pellets was struck at the eastern mint of Constantinopolis. Three nearby mints also struck coins during the early fourth century: Heraclea, Nicomedia, and Cyzicus (figure 324). During 311–335 all three of these mints struck different coins with three pellets in a straight line, and during the same period, no other Roman mint struck coins with three pellets, other than the Labarum coin of nearby Constantinopolis.

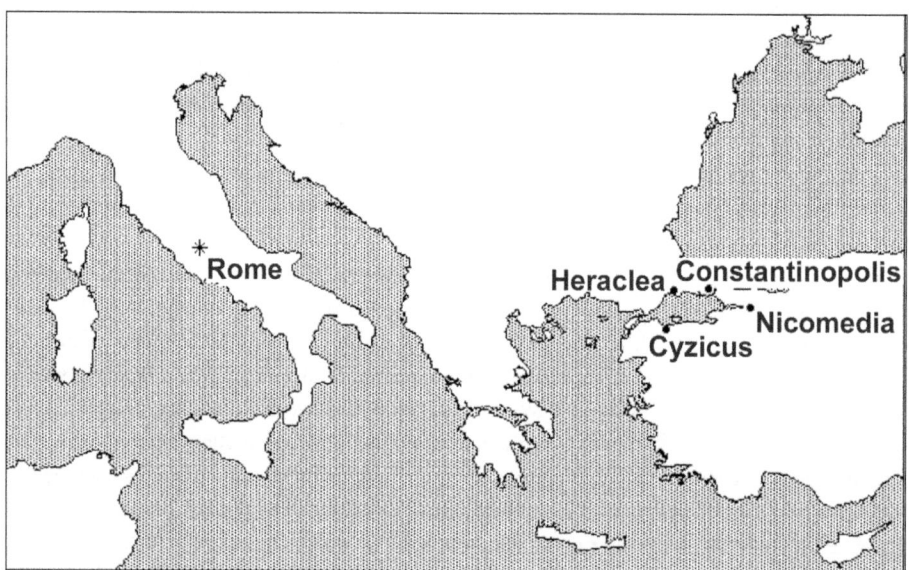

Figure 324. Eastern mints

In Cyzicus during the first half of 311, folles were struck in the names of Galerius, Licinius, Maximinus, Constantine, and Galeria Valeria (daughter of Diocletian) that had three pellets in a straight line in the reverse field of the coins (figures 325–326). Clearly these pellets did not represent Constantine and his sons and these coins predate the Battle of Milvian Bridge.

Is it possible that the three pellets on these folles represent a three-planet conjunction? During the first week of October 310, Mercury, Venus, and Jupiter set together in the west, just after the sun, in a 6° straight line (figure 327).

The three pellets on the Labarum coin can be further explained by examining a series of

6. Visions Explained 115

Figure 325. Licinius follis of Cyzicus

Figure 326. Galerius follis of Cyzicus

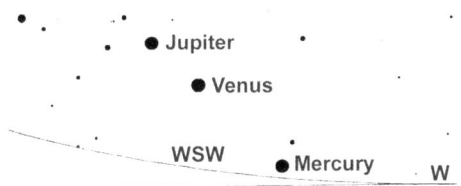

Figure 327. October 6, 310

three coins struck between 330 and 336 to commemorate the founding of Constantinople. The reverse designs on these coins are a she-wolf nursing Romulus and Remus (*URBS ROMA* type), Victory standing upon a ship's prow (*CONSTANTINOPOLIS* type), and two soldiers holding spears and shields with one or two vexilla between them (*GLORIA EXERCITVS* type). This last type was struck in the name of Constantine, Constantine II, Constantius II, Constans, and Delmatius. The Victory on prow and two soldier types recognize the naval and land victories involved in defeating Licinius. The *URBS ROMA* type (figure 328) commemorates the Roman legend of the twins who had been cast onto the bank of the Tiber River by King Amulius, but were found and nurtured by the she-wolf. Romulus would later establish the city of Rome. Above the wolf and twins are two stars.

Figure 328. Romulus and Remus

Perhaps the two stars represent the two capitals of the Roman Empire, Rome and the new city of Constantinople. However, the two stars are more likely to represent the Dioscuri. The Dioscuri above the she-wolf was a common design motif for Roman coins. For example, in 309 Maxentius struck a bronze follis with Castor and Pollux, each holding a scepter and horse, above the she-wolf and twins (figure 329).

Some of the mints struck each of these coin types with additional symbols (figures 330–332), most likely to mark the year or sequence of issue (Failmezger 2002). These additional symbols include a letter; a Christogram or *Chi*; a wreath; a leaf or palm branch; a star; a crescent; and one, three, or four pellets. Presumably each of the special symbols commemorates some part of the events leading up to the founding of Constantinople. For example, a palm branch may

Figure 329. Dioscuri and she-wolf

Figure 330. Branch

Figure 331. Christogram

Figure 332. Wreath

represent peace, and a wreath may symbolize victory. The *Chi* was used interchangeably with the *Chi-Rho* on several coin issues of the period.

All the coins of this series bearing astronomical symbols can be related to the second vision of the cross. In Heraclea (330–333) and Nicomedia (330–335), these coins were struck with three pellets in a vertical line between the stars of the Dioscuri above the she-wolf, to the right of Victory on the prow, and in a horizontal line above two soldiers holding spears and shields with two *vexilla* between them (figures 333–335). There can be little doubt that the three-pellet motif was a significant symbol on this Roman coinage and represented an important event. Certainly the locations of the three pellets on these coins are inappropriate for three portraits, as some have suggested the three pellets on the Labarum coin type represent.

Figure 333. She-wolf

Figure 334. Victory

Figure 335. Soldiers

The three stars *URBS ROMA* type (figure 336) certainly suggests that Sol or Mars was added to the Dioscuri. By 322 the image of Sol as a Constantinian symbol was greatly diminished, so Mars may have been a more likely candidate. The crescent symbol represents the moon (figure 337). During the second vision of the cross, Constantine saw a thin, waning, crescent moon in Taurus next to the celestial *Chi-Rho*. Furthermore, the crescent moon seen during the second vision may have been taken as an omen of a naval victory.

Grant (1993) states that the success of Crispus's naval forces over Licinius's fleet was compared to the Battle of Salamis in 480 B.C., when the Athenians defeated the Persian fleet of Xerxes. A decade before this ancient naval battle, under a thin, waning, crescent moon, the Athenians had been victorious over invading Persians at the Battle of Marathon. The Athenians

Figure 336. Three stars

Figure 337. Crescent

had commemorated the significance of the crescent moon to the victory by adding a crescent above the owl on their coinage.

The one-, three-, and four-pellet symbols can be explained by reconsidering the positions of the planets during the second vision. Mercury, Saturn, and Mars were in a straight line in the constellation of Gemini (figure 338). Venus was in Taurus. The bright stars in Gemini are Castor and Pollux. Thus the three planets are tied to the Dioscuri, and placing the three pellets between the stars on the *URBS ROMA* type indicates the planetary placement in the heavens.

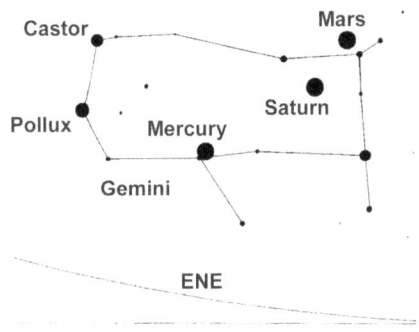

Figure 338. Gemini, July 3, 324

The *URBS ROMA* type with four pellets was struck in Rome, and the pellets are not between the stars on the coin, but rather on the wolf's shoulder (figure 339). Some of the other special symbols were also placed in this location on some of the coins. The *URBS ROMA* type with a single pellet, unless within a wreath, is found not between the stars, but on either side of the wolf or on its shoulder (figure 340). The single pellet may well represent Venus, as this bright member of the conjunction would be incorrectly situated if associated with the stars of Gemini.

Licinius's defeat was the final step in Constantine's achievement of his goal, and the second vision was the omen for the beginning of the end of Licinius. This supernatural atmosphere pervaded the last days of Constantine, who even before his death was accepted into the divine

Figure 339. Four pellets

Figure 340. One pellet

ranks, where, at Jupiter's request, he designated his own son as his successor, and where the heavenly throng accepted his choice.

The symbols on Constantinian coinage build a strong circumstantial case to support the celestial re-creation of the second vision of the cross as presented in this chapter, and its influence on the rise of Christianity in the Roman Empire. This might imply that the second vision may have been more significant to Constantine than the first vision, especially with regard to his unification of the empire and founding of Constantinople. Perhaps the three pellets on his coinage were the basis for the immobilized design that continued on the Christian coinage of Europe for more than a 1,000 years.

The mixed Dioscuri and Christogram on the *URBS ROMA* type can lead to only two possible conclusions. Either Constantine maintained a co-existence of Christian and pagan observances or the *Chi-Rho* did not have a Christian connotation until Eusebius used it as religious propaganda.

Whether or not the three pellets represent a planetary conjunction, this evidence presents a strong argument that the three-pellet motif on the *SPES PVBLICA* Labarum coin is a symbol of the eastern Roman mints and was used on the Labarum as a common device of the region. The Labarum follis is not the first "Christian" coin. It has no more religious significance than the *Chi-Rho* types issued as early as 315.

Under This Sign

Even though Constantine's successors frequently depicted the Christogram on a standard, only a few of his coins showed this motif. On a 336 coin, Constantine is shown in military garb, holding a standard bearing the Christogram (figure 341).

His sons' coins show the Christogram more often. For example, one coin type of Constantine II depicts a Labarum portraying a Christogram on a standard containing three pellets (figure 342), and another coin has a cross with standards of three annulets on each side (figure 343). Although the three pellets or annulets may be purely decorative or represent medallic portraits, rather than a reference to a celestial event, there is certainly a strong relationship with the three pellets on the Labarum depicted on his father's coinage.

Figure 341. Constantine and Christogram

Figure 342. Constantine II (336)

Figure 343. Cross? (334–335)

Of greater importance, coinage struck by Constantine II and Constans clearly reflect a move toward Christian symbolism. The *GLORIA EXERCITVS* coin portraying a cross is particularly significant, as crosses on coins did not become common until the end of the fifth century. Whether the symbol between the two standards is actually a cross or the letter *T* with a pellet above is unclear, but a cross is more likely.

Although some of the coins of Constantius II depict a Christogram, historical records show that he did not hold the same Athanesian Christian views as his brothers. Constantius II was a member of the Arian sect, a Christian following that rejected the Creed of the Trinity.

Constantius II saw another vision of the cross, and symbolic references of this event on some of his coins may add to the discussion of the visions of Constantine. In 350 one of Constan's generals, Magnentius, was declared as Augustus by his troops and had Constans assassinated. Constantius's sister persuaded Vetranio to assume the title of Augustus, because she feared Magnentius, and her brother was occupied in battles in the eastern part of the empire. After his return, Constantius II forced Vetranio to abdicate his position on December 25, 250, and to retire as a private citizen.

After seeing a new vision of the cross, Constantius II was victorious over the usurpers. The actual decisive victory came at Mursa in September 351, although Magnentius was finally defeated and committed suicide in 353. According to John the Monk, the vision was seen over all Jerusalem in the daytime during the festival of Pentecost. Today Pentecost is celebrated on the seventh Sunday after Easter, but traditionally was observed in the spring as the Jewish wheat festival of Shavuot. In both the spring and summer of 351, planetary conjunctions formed a celestial *Chi-Rho* with the planets, although only the spring conjunction involved Mars, and as with Constantine's second vision, Mars was next to the Dioscuri during the heavenly gathering (figure 344).

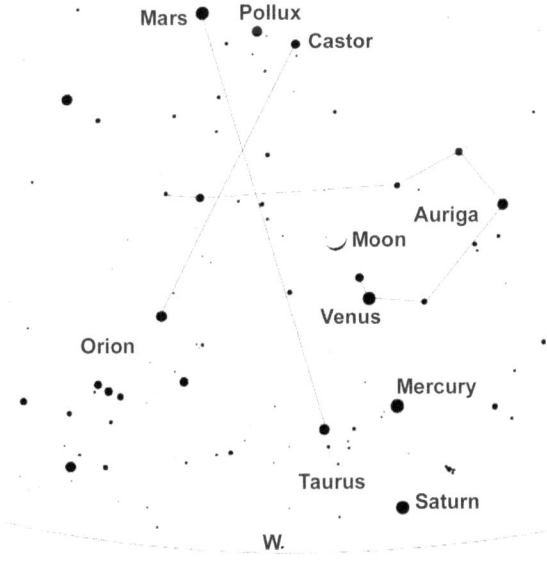

Figure 344. April 15, 351

Constantius II, Vetranio, and Constantius Gallus struck coins showing Victory crowning a soldier holding a standard bearing a *Chi-Rho* symbol. The reverse legend on these coins is

HOC SIGNO VICTOR ERIS (figure 345) and may commemorate Constantine's earlier visions and the new vision. This type of coin was only struck at Siscia and Thessalonica, which were mints under Vetranio's control until his abdication. Most scholars date all of Vetranio's coins to 350, and therefore this type of coin in his name could not have been struck after the April 351 planetary conjunction. Constantius Gallus was not elevated to the rank of Caesar until March 15, 351, so his coins of this type were most likely struck after the celestial event. Those coins in Constantius II's name were probably struck after he took control of Vetranio's mints, and could have been struck before or after the April 351 conjunction. Also Vetranio may have struck coins in both his name and Constantius II's to show his support.

Figure 345. Hoc Signo Victor Eris

There is an explanation that could solve the dilemma of Vetranio's coins being struck prior to the April 351 conjunction. During the fall of 350, Mars was in Gemini and Saturn was in Taurus, and the two planets and bright stars formed a Chi-Rho (figure 346). Perhaps Vetranio struck his *HOC SIGNO VICTOR ERIS* coin to show his allegiance and to honor the house of Constantine, and Constantius II and Constantius Gallus seized a similar opportunity after the April 351 conjunction.

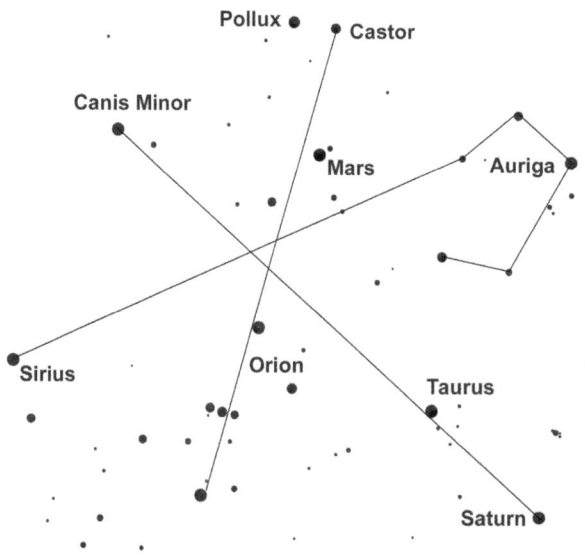

Figure 346. October 13, 350

Paganism and Christianity

The coinage design examples presented in this chapter clearly show that both pagan and Christian symbolism co-existed as the Roman Empire made the transition from its ancients gods to a modern religion. The three pellets, whether a representation of a celestial event as presented here or as a symbol of something else, became part of this religious evolution.

The three-pellet symbol gradually assumed the identity of the Trinity on later European coinage and religious objects. Is it coincidental that the coin depicting three pellets on the Labarum emerged shortly after the concept of the Trinity was officially recognized at the Council of Nicaea in 325? Or was the symbol an intentional device to transfer the concept from pagan gods that ruled the planets to a Christian belief?

Ancient writings often described Christ as a source of light (figure 347). Perhaps this association was a mechanism to transition the worship of the ancient sun god, Sol, to a Christian spiritual entity.

Figure 347. Ancient mosaic showing a Sol-like Christ

The rise of Christianity in Rome and the legends of the visions of the cross were recorded for future generations. The heavenly conjunctions that are the probable basis for each of the visions can be re-created, and symbols on contemporary coins add another dimension to determine what was in the mind of ancient Romans.

7

Norman England

William the Conqueror

Stephen

"The coins are important because they are amongst the most precise evidence we have from medieval England" (King 1994).

Edward the Confessor, the Anglo-Saxon king of England from 1042 to 1066, had promised his distant cousin William, duke of Normandy, succession to the throne of England. On his deathbed, however, Edward named Harold, earl of Wessex, to succeed him. Harold II was crowned on January 16, 1066, but William of Normandy had not forgotten Edward's promise. Indeed, William asserted that the promise of succession had been carried to him by Harold in 1064 and that Harold had sworn an oath of allegiance to him.

On September 27, 1066, William's army of 7,000 men crossed the English Channel, and William and Harold met at the Battle of Hastings on October 14. Harold was killed and England fell under Norman rule. Halley's comet of April 1066 was taken as heralding William's victory. Over the next few years, William the Conqueror and his army marched through England, consolidating his power and building castles to secure his kingdom. William spent 20 years in control of both sides of the Channel and died in 1087 in Normandy, six weeks after falling off his horse.

William's oldest son, Robert, who had rebelled against him, secured control of Normandy, but William had named his second son, Rufus, as his successor in England. William's daughter Adela married Stephen of Blois, whose son Stephen was to rule England half a century later. William's third son, Henry, became king of England when Rufus died in a hunting accident in 1100.

Upon learning of Rufus's death, Henry seized the royal treasury in Winchester and was crowned three days later. His older brother Robert garnered the support of some noblemen, but agreed to remain in Normandy only after negotiating a yearly pension from Henry. Once Henry secured his position in England, he turned his attention to taking Normandy from Robert. In 1106 Henry defeated Robert at the Battle of Tinchebrai and regained control of both sides of the English Channel.

After the death of his only legitimate son, William, in 1120, Henry named his daughter, Matilda, as his successor. Some of the English nobility did not favor the idea of having a queen as the sovereign of England, but Henry's strength forced them to give their allegiance to the throne. Matilda was the widow of Emperor Henry V of Germany. Against her wishes, Henry forced the her to marry Geoffrey of Anjou in 1127, and this marriage of politics was a strained relationship at best. When Henry died on December 1, 1135, Matilda was in Anjou with her husband. Her cousin Stephen of Blois, grandson of William the Conqueror, hurried to England from Boulogne, got the support of many of the noblemen, and was crowned as king of England on December 22, 1135. Some of the nobility saw Stephen's claim to the throne as illegitimate, and minor uprisings occurred.

In 1138 civil war and anarchy broke out in England, when Matilda, with support from her half-brother Robert of Gloucester, claimed her right to the throne. Robert's initial military actions may have been in his own interest to claim more territory during this unsettled period, but he soon pledged his allegiance to Matilda and her cause. Matilda landed in Arundel, England, on September 30, 1139, and stayed there until Stephen's siege on October 15, when Stephen let her go free and his guards escorted her to Gloucester. Matilda then traveled to Bristol to set up her own command.

The first five years of Stephen's reign saw many castle sieges and three major battles. In 1138, at the Battle of the Standard, Stephen's barons defeated an incursion by David I of Scotland, Matilda's uncle, and Stephen entered into an uneasy truce with the Scottish king. On February 2, 1141, Stephen was captured at the Battle of Lincoln by Matilda's supporters and imprisoned at Bristol. The Empress Matilda, however, failed to garner public support for her coronation, and the civil war continued.

During the Battle of Winchester on September 14, 1141, troops loyal to Stephen captured Robert of Gloucester, and Robert and the king were both released in November 1141 as part of the Treaty of Winchester. Matilda's failure to consolidate her power during Stephen's imprisonment was the practical end of her dream to rule England, and the struggle then turned to gaining the kingdom for her eldest son, Henry of Anjou. The next 12 years were filled with unrest and military engagements, and the civil war finally ended in 1153, with Stephen acknowledging Henry as his successor. Henry's reign was the beginning of the Plantagenet line and the end of Norman England. The coins of Norman England are marked by symbols of power and divine omens to assert the strength and legitimacy of the throne during this period of transformation from Anglo-Saxon rule to a new order.

The correlation between astronomical symbols on the Norman coinage of England and actual celestial events is not a new concept. For example, in a series of articles in the *British Numismatic Journal* between 1905 and 1915, Carlyon-Britton detailed the coinage types of William I (the Conqueror) and William II (Rufus). In his articles, he notes several celestial events that could be the basis for the symbols on these coins; however, during the last century such correlations either became unfashionable or were relegated to the status of a curiosity. This chapter will be demonstrate that astronomical symbols on the Norman coinage of England can be used as a valuable tool to help sort the order and date the different types of coins issued.

When considering a series of medieval (and ancient) coins where little was recorded about the numismatic record other than the existence of the coins themselves, both the sequence of the types and the dates when they were struck can be determined by a variety of methods that include knowledge of the minting process, association with other coins and objects discovered in hoards, and correlation of the motifs with known historical events. However, there are usually gaps in the data and disagreements among the researchers.

Hammered coins were almost uniformly struck with the obverse die situated in an anvil-like base and the reverse die embedded in a punch that was struck with a hammer. The reverse die thus received more stress than the obverse die and had to be replaced more frequently. As all the dies were hand made and no two dies are exactly alike, a single type of coin can have different obverse and reverse die combinations. Except for the first set of dies, some coins struck with a new obverse die were usually struck with a reverse die that was also used with the previous obverse die. Within a series of coins of the same type, and given a large number of examples to examine, one can link various obverse and reverse die combinations to recover the sequence in which each coin was struck.

For different coin types one can use two direct approaches to determine sequence. Sometimes older coins were overstruck with new dies of a different type, and if some of the original design is viewable, the overdesign must have been struck at a later date than the underdesign. In addition, sometimes a moneyer created a reverse die punch for a new coin type, but the previous type of obverse die was still in place when coins were struck. A coin with a mixed type of obverse and reverse is known as a mule, and the obverse design almost always preceded the reverse design in sequence. Mules and visibly overstruck coins should have been removed and not placed into circulation, but some examples made it past the inspection process.

When overstruck coins and mules are not available for part of a coin series, researchers order the coin types by systematic changes in coin weight, metal fineness, or design elements and style. In addition, discovered hoards provide evidence for sequence determination if they contain coins of one type but not of another and can be correlated with information or other objects in the hoard for which a date, or range of dates, can be determined.

Determining the date of issuance for each coin type in a series is much more difficult than determining the sequence. Obviously, time was needed to create coin dies when a new sovereign came to power, and the coins of the ruler's predecessor may have been struck for a while after the change in authority. Sometimes new types or minor design changes can be correlated with the opening of new mints or some other political or historical event.

The numismatic community has extensively debated the sequence, and especially the dating, of the Norman coinage of England. The current starting point for sequencing is based on a detailed analysis of mules, overstruck coins, and lettering style that was conducted by Brooke (1916), although more recent evidence has resulted in the reordering of some of the coins of Henry I.

In 1966 Dolley postulated that the late Anglo-Saxon and Norman coin types were issued with new design types on a somewhat regular two- or three-year cycle, tied a few of the coins to dated historical events, and convinced many scholars to adopt his hypothesis. However, Brand (1984), then president of the British Numismatic Society, strongly objected to Dolley's hypothesis as pure conjecture and force fitting of coin types to a regular time schedule that had no basis in fact, and had to be adjusted when the number of types did not evenly fit into the possible range of dates. Brand also suggested that multiple coin types could have been struck at the same time, and that no basis existed for assuming otherwise.

Although the debate continues, there is now general agreement on the sequence of most of the Norman series and some agreement on a range of dates for each type; however, there is no absolute confirmation of what is currently accepted. No correlation of astronomical symbols on these coins with the dates of actual celestial events seems to have been considered since Carlyon-Britton. This chapter will take actual celestial events into account as a tool for sequencing and dating the series.

Anglo-Saxon Precedence

The preservation of local monetary practices was required for economic stability in medieval times. Significant deviations of coinage design might raise into question the value of both circulating and new pieces and could therefore have a negative impact on commerce. Previous chapters have discussed the use of astronomical symbols on Anglo-Saxon and on some Norman coins to represent comets, solar circles, and supernova. To better appreciate the representation of celestial events and transition to Norman coinage, a few more examples of late Anglo-Saxon coins are presented here.

The coinage of Edward the Martyr (975–978) is followed by the first coin of Aethelred II (978–1016) with only a legend change to denote the new king and the addition of three pellets in a line to the left of Aethelred's head (figures 348–349). On September 9, 978, Saturn, Mercury, and Venus rose in the east in a straight line separated by only 1.5°, and the pellets most likely represent this conjunction (figure 350). Subsequent issues of Aethelred II omit the pellets.

Figure 348. Edward the Martyr   Figure 349. Aethelred II

In 1003 Aethelred II issued coinage depicting a helmeted king in armor, perhaps to commemorate his massacre of the Danes in England on November 13, 1002 (figure 351). The reverse

Figure 351. Aethelred II pyramids?

Figure 350. September 9, 978

of design contains a symbol in each quadrant that depicts two lines emanating from three pellets. Perhaps this design is a pyramid symbol representing a comet that was taken as a favorable omen for his victory?

Cnut's second penny type, struck between 1024 and 1030, depicts a helmeted king on the obverse with an annuletted cross with pelleted annulets in each quadrant on the reverse (figure 352). The annulets suggest that an eclipse may have been depicted. At least one variety of Cnut's first issue, struck in Huntingdon between 1017 and 1023, contains annulets (figure 353).

Figure 352. Cnut, second issue   Figure 353. Cnut, first issue

A total solar eclipse crossed over northern England on January 24, 1023, and York was just south of the path of totality. From London, a sliver of the sun remained uncovered (figures 354). Perhaps the rare first issue variety of Cnut with annulets was struck after the eclipse. The annulets on Cnut's second issue are not complete circles, perhaps representing the partially eclipsed solar disk. Consideration of Cnut's Danish coinage appears to confirm this hypothesis.

Coinage was first struck in Denmark, which included southern Sweden and Norway, around 1000, but it was Cnut who fully established coinage for the region. Cnut was in complete control of all of Denmark, Britain, and Ireland, and many of his Danish coins were either imitations of the English types of Aethelred II or were similar to his contemporary coinage in England.

At the Danish mints of Lund and Roskilde, Cnut struck pointed helmet pennies identical to his second English issue. One of the Roskilde helmet pennies was also struck with a different reverse containing a small cross and an annulet (figure 355). In Viborg two types were struck with annulets, and a third penny had four pellets in two of the reverse quadrants and a thick crescent and single pellet in the other quadrants (figure 356). During the eclipse, Saturn was next to the partially eclipsed sun and Mercury, Venus, and Jupiter were low on the horizon (figure 357). The total eclipse of 1023 that crossed northern England, Scotland, and Ireland is most likely the basis for the designs on these coins of Cnut.

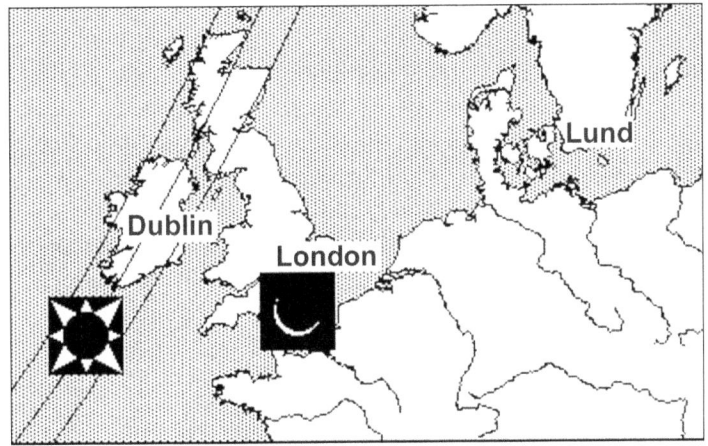

Figure 354. Path of the January 24, 1023, total solar eclipse

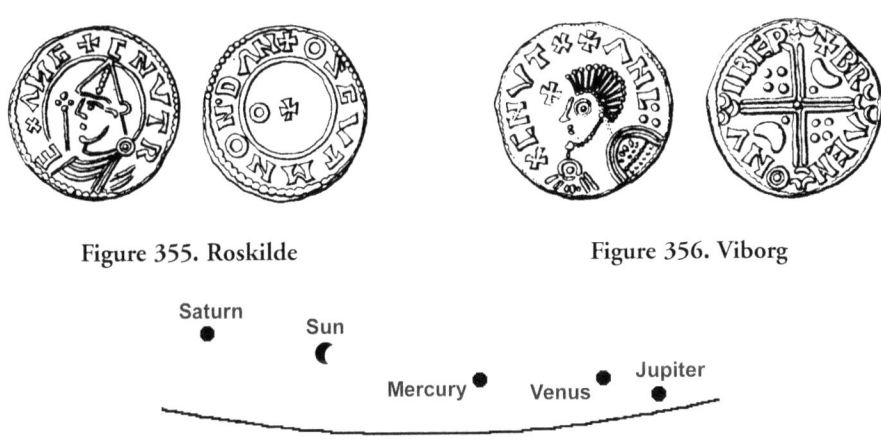

Figure 355. Roskilde   Figure 356. Viborg

Figure 357. January 24, 1023 (Denmark)

The English type of Harthacnut (1035–42), struck in Lund, also had crescents (figure 358), probably in reference to an annular solar eclipse on August 22, 1039. Similar pieces were struck for use in Denmark. This eclipse crossed the southwestern tip of England, would have been seen as a 81 percent partial eclipse in London, and with 50 to 60 percent of the solar disk covered as seen in Lund. Harthacnut was succeeded in Denmark by Magnus the Good (1042–47), who issued coins with crescents on the reverse and an annulet to the left of his bust (figure 359). The annulet may represent the 1039 eclipse.

Figure 358. Harthacnut (Lund mint)   Figure 359. Annulet/crescent type of Magnus

The Coinage of York

Vikings invaded England late in 865 and initially captured York in 866. Then the Norse invaders moved into East Anglia in 869 where they put King Edmund to death. By the end of the ninth century, however, the Viking settlers considered Edmund a saint. At least one type of their St. Edmund memorial coinage, struck between 895 and 910, contains annulets (figure 360). On January 1, 865, a total solar eclipse crossed England and just to the west of Denmark, and may have been taken as an omen for the invasion later that year. On October 29, 878, both England and Denmark were in the path of a total eclipse. Either of these celestial events may be depicted on Viking coinage and may have remained as an immobilized motif.

Figure 360. St. Edmund memorial coinage

According to Pegge (1772), the annulet was not an ancient symbol of York. During most of the first half of the 10th century, Norse kings of Dublin also ruled in York. Although annulets are sometimes found on English coinage in the 9th and 10th centuries, the sword was considered to be the symbol of York under Viking rule. Beginning with the invader Sihtric in 921, and continuing through the reign of Eric Bloodaxe in 954, many coins of York prominently displayed a sword.

A few of the Viking types have crescents or annulets. A rare variety of Sihtric presents a dilemma. On the obverse side of the coin the Viking sword and Thor's hammer depict pagan beliefs. On the reverse a cross represents Christianity (figure 361). The depiction of both pagan and Christian symbols on the coin is not unusual, but the annulets on the reverse are interesting. Sihtric died in 927 and his successor, Guthfrith, was ousted by Aethelstan of Wessex the same year. The Viking rule of York resumed in 939 when Anlaf Guthfrithsson recaptured the city.

Figure 361. Sword type of Sihtric (York)

The only solar eclipses during the first half of the 10th century that might be depicted by the annulets are a total solar eclipse in 912 that crossed to the west of England, and an annular eclipse in 934 that crossed just south of both England and Denmark. Neither celestial event occurred during the reign of Sihtric. Perhaps the annulet was immobilized from an earlier time. A few of the late ninth century Danish coins depict a circle of pellets; some with a central pellet suggestive of a solar symbol (figure 362).

The first English use of an annulet on York coinage was by Eadgar (959–975). These coins

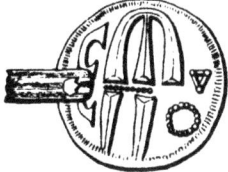

Figure 362. Late ninth century Danish types

are found with annulets, crescents, or clusters of three or seven pellets in the designs or legends (figure 363). On May 17, 961, an annular solar eclipse crossed the northern coast of France, and was seen in England as a very thin crescent. In London, 91 percent of the solar disk was covered, and in York, an 87 percent partial eclipse was seen. Eadgar had unified all of England by 959, and perhaps this eclipse was taken as a divine omen. Although a few issues of Edward the Martyr (975–978) and Aethelred (978–1016) have annulets, the annulet motif did not become predominant on Anglo-Saxon coinage until the second issue of Cnut in 1024.

Figure 363. Annulet type of Eadgar

The coinage of Edward the Confessor (1042–66) has a variety of designs. Many of his coins have an annuletted cross on the reverse, possibly a continuation of the style of Cnut. Coins of the York mint have an annulet in one of the reverse quadrants (figure 364), which perhaps remained as an immobilized symbol of the 1023 or 1039 eclipses.

Figure 364. Edward the Confessor (York)

The adoption of the annulet as a mint mark of York for coins of Edward the Confessor is of considerable interest. During his reign, only the coinage struck at York exhibited a consistent mint mark, but what event in the history of York would be so important that the annulet was used as its mint mark? Perhaps the annulet mint mark represents the 865 eclipse that may have coincided with the Viking invasion, or perhaps the solar eclipse of 1023 may yield an explanation. In that year on May 28, Wulfstan II, who was archbishop of York, died. Was the eclipse taken as an omen for his death or a commemoration of his service?

Anglo-Saxon to Norman Transition

There is no question that astronomical symbols were used on late Anglo-Saxon coinage. Annulets, crescents, pellets, pyramids, and even occasional stars are found. The use of such symbols on Norman coinage was thus a continuation of the local and customary practices that had preceded William I's conquest of England.

According to Carlyon-Britton (1905), Theoderic designed the last few coinage types (figures 365–366) for Edward the Confessor, the penny of Harold II, and the first five types of William I. Thus one would expect his style and motifs to bridge the coinage of all three kings. Theoderic was succeeded by Otho, who introduced a large reverse cross motif that persisted until the last type struck for William II. Otho's son, Otho FitzOtho, formally succeeded his father by means of a charter granted in 1101, but similarities between the last type of William II and the first type of Henry I indicate he may have begun work prior to his formal appointment.

Figure 365. Last coinage of Edward the Confessor (1064–66)

Figure 366. Penny of Harold II (1066)

The Coinage of William I and William II

Thirteen different types of pennies were struck during the reigns of William I (the Conqueror) and William II (Rufus) between 1066 and 1100. No information on these pennies directly identifies under which William each was struck, but with one possible exception, the sequence of pennies can be determined through analysis of mules, overstrikes, and design changes.

Numismatists customarily assign a type number to the Norman coins based on Brooke's analyses and numbering designations. Assuming a somewhat regular interval between the issuance of new types and bearing in mind the length of each William's reign, the first eight types are assigned to William I and the last five to William II, although opinions differ about whether type VIII belongs to William I or II, or possibly it spans the reigns of both kings.

The first type of William I features a left-facing profile and a sceptre on the obverse and a cross fleury on the reverse (figure 367). Carlyon-Britton asserted that the central annulet of the cross fleury represents the ring given to the king during his coronation. In 1064 an annular solar eclipse crossed Scotland and just to the north of Lund, Denmark. Danish coins of Sven Estrithson depict a

Figure 367. William I, type I

crescent and annulet, most likely in reference to this celestial event (see figure 240). However, this eclipse was not recorded in European chronicles, and a central annulet had been a common design feature on Anglo-Saxon coinage since the time of Cnut, and therefore must be considered as an immobilized design element. A few very rare varieties of William's first type omit the sceptre, and extremely rare varieties are found with the bust facing to the right. The type I pennies of William were issued soon after his victory at the Battle of Hastings, either late in 1066 or early in 1067.

William's type II reverse has pyramids in each quadrant oriented to form a star pattern (figure 368). The pyramids are similar to those of the last coin of Edward the Confessor, on which the comet of 1063 or 1065 may have been represented, and Halley's comet of 1066 is most likely depicted on this type II penny. The obverse depicts the king wearing a bonnet-like crown. As several type I/II and II/III mules have been found, the sequence of these three coins is not in doubt, but the type II penny may have been introduced earlier than the currently accepted date of 1068 shown in North's standard reference (1994).

Figure 368. William I, type II

One might ask why the comet of 1066, taken as an omen for William's victory, was not depicted on his first coin type. The first two to three years after the conquest were marked by various military campaigns to secure England under Norman control. William most likely ordered coinage to be struck in his name as soon as possible, and existing dies and those in production for Harold II under Theoderic's direction may have been modified with William's name and used to strike coins until a new design and dies could be fabricated.

Harold had been crowned as king on January 6, 1066, and was killed at the Battle of Hastings on October 14 that year. Certainly it took some time to have dies made for Harold, and additional dies would have been in production at the time of his death. The profile bust and sceptre design of William's first type is almost identical to that on the penny of Harold except for the absence of a beard on William's bust. A few pennies struck during Harold's brief reign are also found without a sceptre or with a right-facing bust, adding to the evidence that existing or partially completed obverse dies may have been used for William's first type. However, this does not explain the cross fleury reverse instead of the pyramids reverse on the type I penny if the comet of 1066 were to be depicted to show divine intervention.

Examination of a few mules that have the obverse of the last type of Edward the Confessor with reverses from either the first or second type of William may hold the answer. These mules are extremely unusual, as mints where these coins were issued also struck coins of Harold, and mules almost always represent the next coin in sequence. One such mule struck in Lincoln has a William type I cross fleury reverse. A mule thought to be from Shrewsbury and another with a blundered (non-meaningful) reverse legend have William type II pyramid reverses. The blundered legend penny may be either a contemporary forgery or was struck by a moneyer who did not wish to have his name associated with the coin. These mules suggest that perhaps both the type I cross fleury and type II pyramid reverse dies of William were produced at about the same time, and possibly the type I reverse die was made during the reign of Harold.

Although no direct evidence supports the following, a possibility is that the design and production of dies for Harold's coins had originally been intended as the next issue for Edward the Confessor, as the pyramid type had already been in circulation for two years. New reverse dies with a simple reverse *PAX* (peace) may have been created quickly upon Edward's death. However, production may have started on the cross fleury reverse dies in 1066 that were being saved for a future second coinage of Harold, but were then used on the first type of William.

One could argue that the Confessor profile obverse may have been accidently confused with the William type I profile obverse when the mule from Lincoln was struck, even the busts are facing in opposite directions. With the Norman invasion, the resistance by Anglo-Saxon nobility, and the production of coinage from three different kings in the same year or so, mint operations may have been in a state of confusion and haste, and minting errors are understandable. A mule of a William type I obverse with an Edward the Confessor hammer cross reverse (c.1059–62) was struck in London, thereby underscoring the confusion at the mints during the type I coinage production.

However, the Confessor and type II reverse mules are a different matter if type II pennies were not issued until 1068. The type II obverse has a facing bust and should have been easily distinguished from a profile bust in normal mint operations. More likely the type II penny was authorized soon after the type I penny was first struck at Shrewsbury, both reverse type dies were on hand at the mint, and both the Confessor obverse and the type II reverse dies were used in error. However, an obverse die of Edward the Confessor is unlikely to have been readily available for accidental use more than two years after his death, suggesting a date earlier than 1068 for the striking of this mule.

Furthermore, with regard to the Confessor and type II mule with a blundered reverse legend, a forger is unlikely to have copied an erroneously struck coin that might be more easily detected than a correctly struck coin. More likely a blundered legend die was produced by a moneyer who was unsure if the Norman reign would persist and therefore did not want to have shown allegiance to William. Striking coins with blundered legends for precisely this purpose was common practice later during the anarchy at the time of Stephen and Matilda. Although William did campaign throughout England during 1068–70, this activity was more for castle building to maintain his position than to enlarge the area of Norman control (Beeler 1966). Therefore this mule may have been struck in 1067 shortly after new territory came under William's rule.

Thus, the type II penny of William could have first been struck sometime in 1067, and the type I penny could have been a transitional issue using dies that were not originally intended for William. Under this scenario, Halley's comet of 1066 may then have been depicted on the first coin designed for the new Norman king.

William's type III canopy bust and cross fleury with an annulet in quadrilateral reverse (figure 369) is thought to have been struck between 1070 and 1072 (North 1994). The circular

Figure 369. William I, type III   Figure 370. William I, type IV

feature in the portico above the bust is thought to be a window, and the central annulet on the reverse remains as an immobilized design. During Easter 1070, William was crowned in Winchester by three legates sent by Pope Alexander, and this penny may represent that ceremony. The type IV penny features the king with two sceptres on the obverse, and a central pellet with a cross fleury over a saltire bottonée on the reverse (figure 370).

Figure 371. William I, type V

William's type V penny is extremely interesting. The obverse features a facing bust with a star on each side (figure 371). The reverse has a central annulet with cross bottonée and a return to pyramids in all four quadrants to form a quadrilateral. On a few examples, the quadrilateral shape is more star-like as designed for the type II reverse. Boon (1988) suggests that the two stars represent Normandy and England, but this idea is not supported. William is unlikely to have waited for his fifth coin type to depict the two territories and subsequent types of William II have only one star.

North dates this coin to the years 1074–76, but the stars and pyramids suggest a later date. Carlyon-Britton suggested that the coin was issued late in 1077. His reasoning was that the stars represent two comets, one being Halley's of 1066 and the other being a star that appeared on Palm Sunday of 1077 and was referred to in the *Annales de Waverleia*. Carlyon-Britton stated that 1077 was a troubled time for William, and that he used the new comet as propaganda to create the impression of success.

The reference to comets is likely, and this penny depicts them both as stars and as pyramids; however, the 1077 date is probably incorrect. The reference to a bright star in 1077 being a comet may not be accurate, as no other reliable sources refer to a comet in that year. Perhaps a brilliant meteor was seen on that day.

Although unrecorded in European chronicles, oriental records mention comets in 1072, 1073, and 1074 and a fairly bright comet in 1075. The 1075 comet was recorded in China, Japan, and Korea. It first appeared on November 17 as a bluish-white, Saturn-like star whose tail grew to 10° by November 20, and then was visible for another eight days.

William had gone to Normandy in 1075, and in his absence the first of a series of revolts by English barons occurred when Roger, earl of Hereford, and Ralph, earl of Norfolk, plotted to overthrow the king. Bishops Odo of Bayeaux and Geoffrey of Coutances raised an Anglo-Norman army, and with the help of royal officers successfully put down the revolt prior to

William's return to England in autumn 1075. As in 1066, an autumn comet may have been associated with a victory for William and a crossing of the Channel from Normandy to England. Certainly William desired not only to proclaim his victory over the revolt, but also to re-affirm his divine right to the throne.

If the stars represent the comets of 1066 and 1075, the type V penny would have been issued late in 1075, or more likely, early in 1076. Assuming that this is correct, types III and IV were then stuck between Easter 1070 and early 1076, making each a three-year issue rather than two-year issues.

The type VI penny of William features the king with a sword and a reverse cross pattée introduced by Otho (figure 372). The cross is superimposed over a quadrilateral fleury. Type VII has William facing right with a sceptre and the reverse cross with fleurs in the quadrants (figure 373). The schema proposed by Dolley forced the first eight types of the William pennies into the Conqueror's reign, and artificially assigned two years each to the first four types and three years each to the last four types. Thus types VI and VII are assigned to 1077–80 and 1080–83, respectively. However, if type V were issued early in 1076 after the appearance of the comet, then it may have been struck at the mints until 1078 or 1079, pushing back the dates for types VI and VII by two years.

Figure 372. William I, type VI   Figure 373. William I, type VII

Carlyon-Britton suggested that type VI was struck between 1080 and 1083. He cited a portion of Orderic's account of a speech given by William stating that his foes got nothing but wounds from his sword, and ties this speech to the coin motif. Carlyon-Britton then assigned type VII to 1083–86.

Figure 374. William I and William II, type VIII

The type VIII *PAXS* penny (figure 374) is by far the most common of the entire series of 13 coins, and this suggests that it was minted for a longer period of time than the other coins in the series. Dolley's schema restricted type VIII to the reign of William I with a minting period of 1083–86. Metcalf (1988) argued that *PAXS*-type coins are associated with the reign of a new king, and that the type VIII penny belongs to William II beginning in 1087. Acceptance of type V being struck between 1076 and 1079 would place the introduction of type VIII around 1085, and the type VIII coinage would extend into the reign of William II, perhaps until 1088, if types VI and VII were struck for three years each.

Type VII, however, is the rarest of the first eight types, suggesting a shorter period for its issuance. If one assumes two years for type VII and four years for type VIII to account for the number of coins of each type known, then the years 1082–84 and 1084–88 can be associated with each type respectively. Therefore, type VIII would span the reigns of William I and II. Carlyon-Britton suggested that type VIII was struck for only one year until the death of William I in 1087, but this shortened period would not account for the large number of coins of this type.

The remaining five types in this series were struck during the reign of William II, and two of them have astronomical symbols that can be used to ascertain the dates of issuance.

The type I penny has a profile bust facing right on the obverse and an annuletted cross pattée over cross fleury on the reverse (figure 375). North follows Dolley's schema and assigns the years 1086–89 for this coin, but based upon the previous discussion, 1088–91 may be a more likely period. Furthermore, Dolley's schema would have placed the coin in the reigns of both William I and William II.

Figure 375. William II, type I   Figure 376. William II, type II

The type II penny features a facing bust with sword and a cross pattée within a quatrefoil on the reverse (figure 376). North and Dolley assign this type to 1089–92, while Carlyon-Britton suggested 1090–93. However, in keeping with the hypotheses presented here, 1091–94 would be more likely if type I had been struck for three years.

At this juncture, the sequence of types becomes somewhat unclear. First consider the currently accepted sequence proposed by Brooke in 1916 and adopted thereafter. Type III depicts William II flanked by a star on each side of his bust and annulets at his neck (figure 377). The reverse has a voided cross pattée with a central annulet over a cross annulettée. A type III variety was struck in Dover, Hastings, Northampton, and Norwich without the stars and annulets (figure 378). Dolley assigned this type to 1092–95.

Figure 377. William II, type III   Figure 378. William II, type III without stars

The type IV penny of William II has an obverse that is almost identical to that of type II, but with coarser workmanship (figure 379). The reverse of type IV has a cross pattée over a cross fleury. Dolley assigned the type IV penny to 1095–98.

However, Carlyon-Britton placed the type III penny as the fourth type of William II (1096–99), with Brooke's type IV (1093–96) preceding the two stars type. His justification was threefold. First, he considered that the similar obverse design indicates just a decline in workmanship,

Figure 379. William II, type IV

but explained only that the reverse is similar to that of type I. Second, he cited a coin from his collection with the type IV design overstruck on a type II penny, and assumed not only order of sequence, but also subsequence. Finally, Carlyon-Britton asserted that the comet of 1097 was the basis for the stars of the type III obverse and that the varieties without the stars were struck in 1096 prior to the comet's appearance.

Brooke's analyses and order of type sequencing was conducted a decade after Carlyon-Britton's article. A single coin from Thetford that has a type IV obverse overstruck on a type III obverse and has a type IV reverse is apparently the basis for his sequencing of these two types. This coin is pictured in Brooke's first volume, and it is unclear to this author which obverse was overstruck on the other based on the printed image. However, the type IV reverse would indicate that Brooke's sequence is correct.

The date of issue of these two coins is another matter. Under Dolley's schema the two stars type was struck between 1092 and 1095, and therefore could not represent a comet visible in 1097. As stated earlier, better dates for type II would be 1091–94, and thus type III would have been produced in 1094–96 or 1097 depending on whether this coin was struck for two or three years. These dates precede the comet of 1097 as well. Another explanation for the stars and annuletted cross exists.

On September 23, 1093, the path of an annular solar eclipse crossed eastern Europe and was seen as an 81 percent partial eclipse in England (figure 380). Florence of Worcester and Simeon of Durham recorded identical text in reference to the eclipse in their chronicles: "A very wonderful sign appeared in the sun." The stars and annulets on the type III penny may be a representation of this eclipse. Perhaps dies without the stars used at the four mints were produced before the eclipse, as some pennies from Hastings and Norwich do have the stars.

We now have type IV being issued in 1096 or 1097. Given the approximately equal rarity of types IV and V coupled with the death of William II in 1100, dates for types IV and V of 1096–98 and 1098–1100, respectively, seem appropriate. Type V has the king's bust with a single star and annulet on the obverse, and an annuletted cross fleury and pyramids on the reverse (figure 381).

The star and annulet may be immobilized from the type III penny, but the reverse pyramids are most likely a reference to the comet of October 1097. This comet had particular value as a propaganda device. As Crusaders marched into Syria in 1097, this sword-tailed comet was seen in the heavens and was taken as a divine omen for victory. Inclusion of the comet's representation on coinage struck soon afterward would have reinforced William's authority.

Although the coinage of William II may have exploited celestial events as propaganda devices, some evidence indicates that he did not believe in omens. According to Gerald of Wales (c. 1180–1220), when William threatened to invade Ireland, Murdhac, the lord of Leinster, replied: "Since that man trusts so much to human power and not to divine, I have no fear of his coming." Furthermore, Johnson (1877) states that upon being told of a vision that predicted fatal dangers with regard to the hunting trip during which he was killed, William is reported to

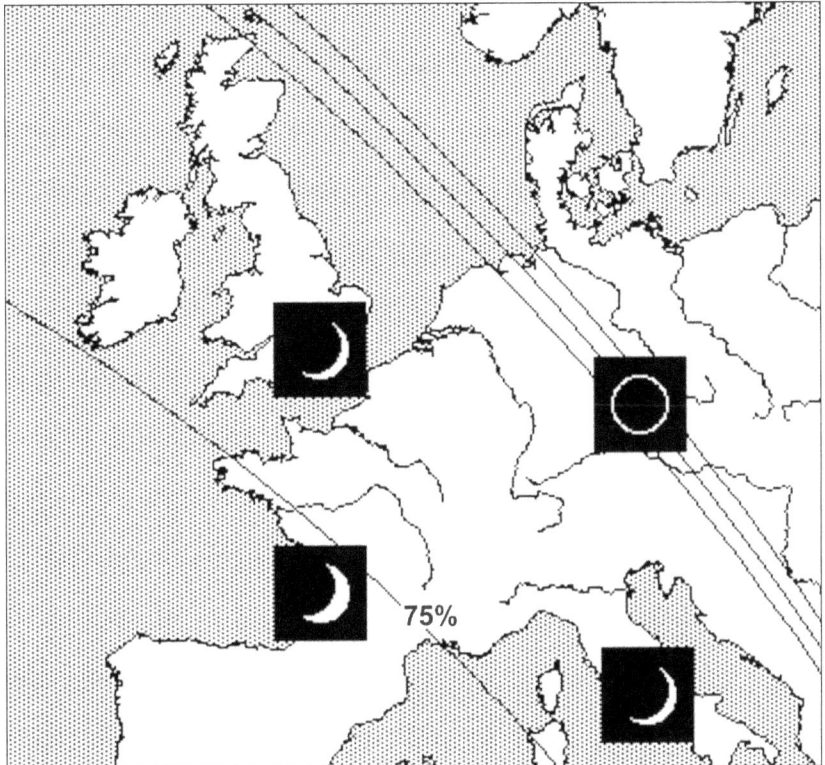

Figure 380. Path of the September 23, 1093, annular eclipse

Figure 381. William II, type V

have said: "Do they think me an Englishman to put off a journey for an old wife's fancy?" Perhaps William's personal belief in physical superiority, rather than divine authority, is reflected on his coinage, because three of his five types depict him with a sword.

The preceding discussion offers a revised dating schema for the coinage of William I and William II based not only on political and religious events, but also on celestial events. These assertions appear to offer a consistent and logical structure for the dating of the coins, and table 7.1 provides a summary, including the precipitating events for some of the designs.

The Coinage of Henry I

The mints struck 15 types of pennies during the reign of Henry I between 1100 and 1135. The inspection of coinage at the mints was much better than during the previous two reigns, and only four known mule combinations escaped detection and were released. Mules link types V/VI, IX/X, XI/X, and XIII/XIV. In addition, a type XV is muled with the first type of Stephen, initially struck in 1136. With only a limited number of mules to consider, Brooke (1916) developed a sequence for the 15 types that also included a grouping of types and their sequence based on epigraphy and on types found in coin hoards. The overlapping of letter styles, coin hoards, and known mules provided a good framework for the proposed sequence.

Table 7.1
Revised Dates for Coins of William I and William II

Type	Description	Dates	Precipitating event
I–I	Profile left Cross fleury	1066–67	Battle of Hastings, October 14, 1066
I–II	Bonnet Cross pyramids	1067–70	Halley's comet, April–June 1066
I–III	Canopy Quadrilateral fleury	1070–73	Easter coronation, Winchester, 1070
I–IV	Two sceptres Cross fleury, saltire	1073–76	Unspecified
I–V	Two stars Cross pyramids	1076–79	Comet, November 1075
I–VI	Sword Quadrilateral fleury	1079–82	Unspecified
I–VII	Profile right Cross fleurs	1082–84	Unspecified
I–VIII	Facing bust PAXS	1084–88	Unspecified
II–I	Profile right Cross, cross fleury	1088–91	Unspecified
II–II	Sword Cross quatrefoil	1091–94	Unspecified
II–III	Two stars/annulets Cross, cross annulettée	1094–96	Annular eclipse, September 1093
II–IV	Sword Cross, cross fleury	1096–98	Unspecified
II–V	Bust with star Cross pyramids	1098–1100	Comet, October 1097

Considerable research conducted since the sequence Brooke suggested has re-ordered some of the types, although numismatists continue to use his type numbers. There is general agreement on the sequence for types I–VI and XII–XV, with the possible exception of the order for types II and III. The order of types VII–XI remains controversial, however. The latest research by Blackburn (1991) suggests a sequence of I–VI, IX, VII, VIII, XI, X, and XII–XV. He estimates dates for each type with a margin of error of plus or minus three years.

In addition to the death of William II in 1100 and of Henry I in 1135, two numismatic events are used to help sort out the dates for the types. In 1107 or 1108, the silver content of the coinage had become so debased, and so many forgeries were in circulation that a decree was issued ordering mints to snick all coins before releasing them to show the internal metal. Some of the type VI coins, and almost all the type VII–XII pennies, were cut following this decree. Thus we know that some of the type VI coins were struck before 1108 and the remainder were struck after the decree.

Debased coinage persisted, however, and the king summoned all of England's moneyers to the Assize of Winchester in December 1124, and had their hands and genitals mutilated as punishment. Apparently a few of them escaped this plight, perhaps by paying a fine. North (1994) asserts that the considerable reduction in the number of moneyers and mints for type XV was a direct result of the assize, and therefore dates type XV to 1125–35.

Celestial events may also hold some of the answers for the design types. For example, comets were recorded in English chronicles in 1106, 1109, 1110, 1114, 1119, and 1132. Other comets were recorded elsewhere in Europe in 1126 and in oriental records in 1113, 1118, 1123, 1127, 1130, and 1131. Of all the comets recorded during Henry's reign, those of 1106, 1110, 1126, 1127, and 1132 were the most impressive in appearance. Solar and planetary events were also recorded in England during his reign, and these events can help support the dating of Henry's coins.

First consider the six types struck prior to the decree to snick the coins (figures 382–387). The coins are presented in the sequence suggested by Brooke and with the date of issue suggested by North. The predominant celestial symbol on types I, III, and IV is the annulet. Types I and

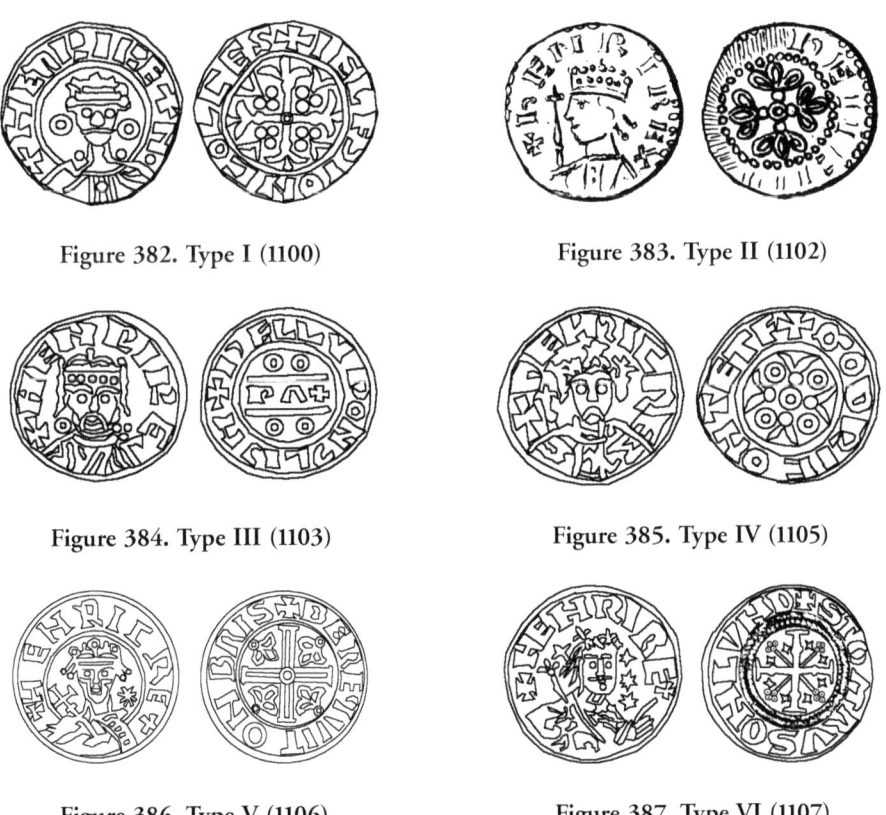

Figure 382. Type I (1100)

Figure 383. Type II (1102)

Figure 384. Type III (1103)

Figure 385. Type IV (1105)

Figure 386. Type V (1106)

Figure 387. Type VI (1107)

IV also depict pyramid shapes. One possibility is that the annulets and pyramids are immobilized design elements from the pennies of William II and represent the 1093 annular eclipse and the 1097 comet, but other possibilities must be considered.

On August 22, 1101, Venus and Jupiter were separated by slightly less than 1/20 of a degree (0.05°). Because of their brightness, they probably appeared as a magnificent, single, bright object. One month later, on September 26, 1101, Venus and Mercury were about 1/60 of a degree (0.016°) apart, and to observers it may well have appeared that Venus had consumed its lesser bright companion. These two events might easily have been used by Henry as a divine sign for his seizing the throne, and perhaps the annulets represent the planetary conjunctions. However, this would be an unusual use of the annulet symbol, and would require that type I was struck late in 1101 after the conjunctions.

An explanation for the four annulets on the reverse of type III may be found in *The Anglo-Saxon Chronicle*, in which rings around the sun, probably due to multiple reflections of sunlight off high-altitude ice crystals, were observed in 1104: "This year Whit-Sunday was on 5 June, and on the Tuesday after there appeared at midday four circles all round the sun, of white colour, each intertwined under the other as if they were painted. All who saw them were astonished for they did not remember anything like it before."

If the four annulets on type III are a reference to the solar circles, then type III was issued after the celestial event in 1104 rather than 1103. Type V (1106) features a star to the right of the king and type VI (1107) has three stars on the obverse and stars in the angles of the reverse. The year 1106 was extremely important for Henry, because that was the year he invaded Normandy and defeated his brother Robert at the Battle of Tinchebrai on the Michaelmas eve (September 29), and in doing so secured his title to the throne of England and once again united England and Normandy.

The year 1106 was a significant one for celestial events as well. In the early mornings of January and February 1106, Venus, Mars, Jupiter, and Saturn formed various configurations within a 10° circle (figure 388). While a 10° conjunction is not extremely tight, it would be astrologically significant. On January 22, 1106, these four planets rose in the east in a diamond shape in the early dawn sky.

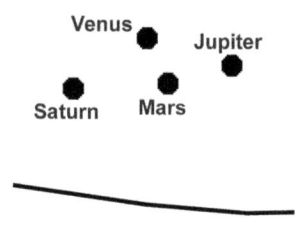

Figure 388. January 22, 1106

The comet seen in February 1106 was spectacular, and the comet, and perhaps the planetary conjunctions, were noted in the *Anglo-Saxon Chronicle*:

> In the first week of Lent, on the Friday, 16 February, in the evening, there appeared an unusual star [comet], and for a long time, after that it was seen shining for a while every evening. This star appeared in the south-west; it seemed small and dark. The ray that shone from it, however, was very bright, and seemed to be like an immense beam shining north-east; and one evening it appeared as if this beam were forking into many rays towards the star from an opposite direction. Some said that at this time they saw more strange stars [the planetary conjunctions?]. However, we do not write of it more plainly because we did not see it ourselves.

In the *Chronicle of the Counts of Anjou*, John, a monk of Marmoutier Abbey, confirms the unusual star of 1106 as a comet:

> In the year of our Lord 1106, there was a period of forty days when an ever-growing comet appeared every evening and filled the world with amazement. Casting its rays of shining splendor against the misty sunset, it seemed more fiery initially and then became less clear, gradually burning out until after forty days it disappeared altogether, or so they say.

The star to the right of Henry's bust on type V is surely a representation of this comet, and thus verifies the suggested date of 1106. The three stars on type VI must then represent the three comets associated with Norman victories, that of William I in 1066 and 1075 and Henry's victory over Robert at Tinchebrai, and thus type VI would have been struck late in 1106, or more likely in 1107. The type VI penny is extremely rare, and one variety has two stars to the right of Henry's bust with the third star at the end of *REX* in the legend, perhaps signifying that one of the stars belongs to Henry. Another variety adds a fourth star at the end of *REX* and may be reemphasizing his divine right to the throne.

The second group of Henry's pennies consists of those commonly found with a snick in the coin to verify their metal content. The coins are presented in the sequence suggested by Blackburn and with the date of issue suggested by North (figures 389–394).

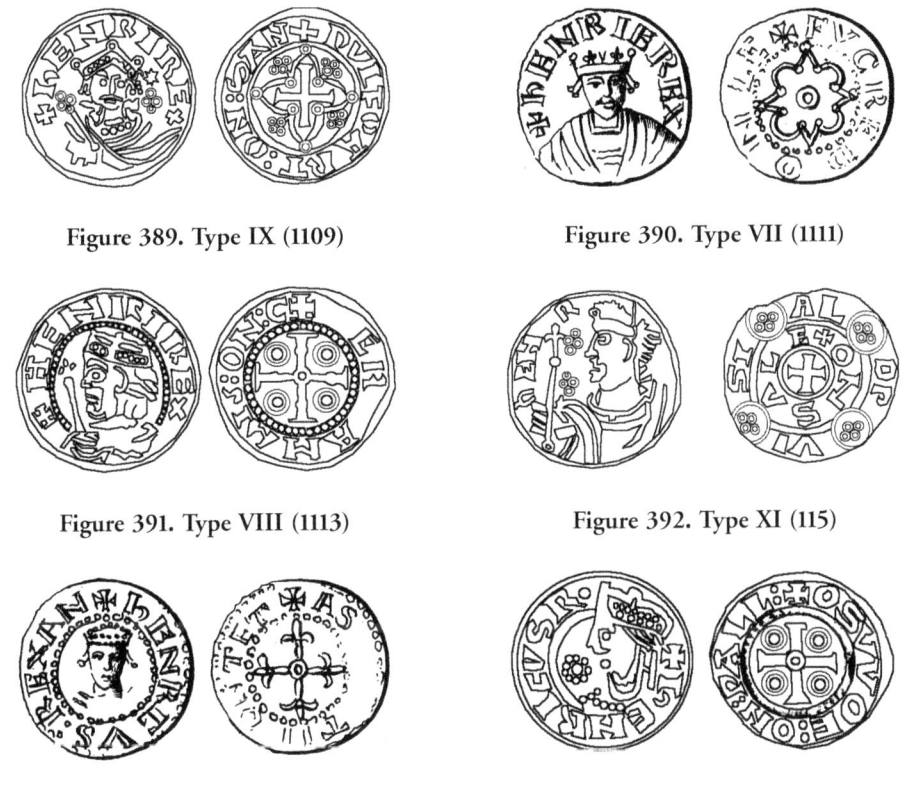

Figure 389. Type IX (1109)

Figure 390. Type VII (1111)

Figure 391. Type VIII (1113)

Figure 392. Type XI (115)

Figure 393. Type X (1117)

Figure 394. Type XII (1119)

The type IX penny has a star and four annulets in a diamond cluster to the right of the bust. The star may be immobilized from the previous two types and represent the comet of 1106, or may depict either of the comets of December 1109 or June 1110, both of which were recorded in English chronicles. The 1110 comet was visually more spectacular, and thus is the more likely choice for this coin, thereby moving its issue date from 1109 to 1110. The four annulets, also depicted on types VIII, XI, and XII, may represent the 1106 planetary conjunction, or possibly the four solar circles seen in 1104. If these annulets do represent planets, then perhaps the annulets on types I, III, and IV do as well. The reverse of type VII features a central annulet and a quatrefoil with pyramids. Although not a stellar depiction, the pyramids may represent the comet seen in May 1114, thereby moving type VII to that year.

Now consider type XII. North dates this coin, the last of the group that was snicked, to 1119. If type XV were issued in 1125, all the types from IX through XIV have to be dated between 1109 and 1124, and a two-year cycle was chosen. This schema also requires that the last type of Henry was struck for a 10-year period. The interesting feature on type XII is a rosette in front of the king's bust. This rosette is perhaps a solar symbol, and as stars had been used on earlier types of Henry's pennies to represent comets, a variation on the stellar motif may have been used to represent the sun. The interchangeable use of rosettes and mullets will be shown later on some coins of Stephen.

On August 11, 1124, the path of a total solar eclipse crossed Norway and Sweden (figure 395) and was recorded in English chronicles as looking "like a new moon," thereby verifying that the crescent of the thin partial eclipse was seen there.

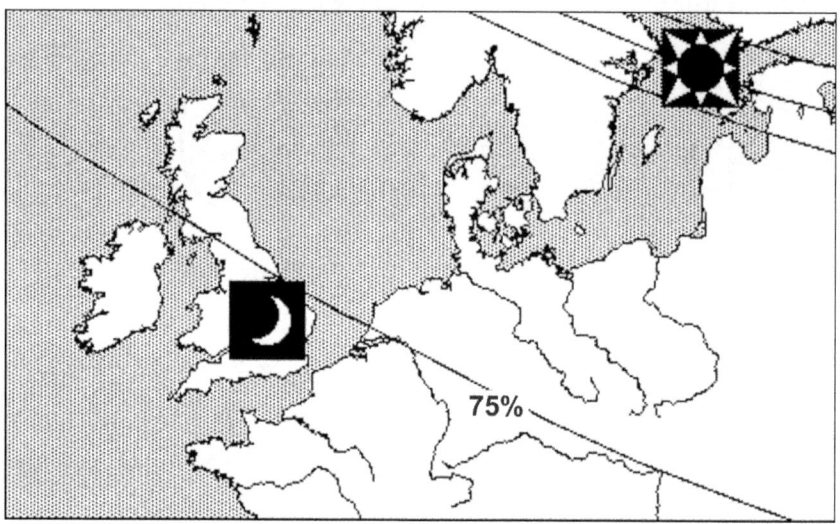

Figure 395. Path of the August 11, 1124, total eclipse

If this eclipse is represented by the rosette on type XII, then this type would have been issued in 1124 just prior to the Assize of Winchester in December of that year. This would imply that types VIII, XI, and X were issued in 1117, 1119, and 1121, respectively, assuming a somewhat equal time between issues. This would also mean that all the pennies struck between the decree of 1108 and the assize in 1124 were snicked, and that types issued after the assize were not, as those moneyers found guilty of debasing their coinage were no longer able to strike coins. The few type XII coins struck prior to the assize would have been snicked, and coins of type XII struck after the assize may have been snicked until the decree was formally lifted and the king had ascertained that the continuing and new moneyers were striking coins of the proper fineness and weight.

The assertion that the number of mints issuing coins would have dropped significantly after the assize is a valid point, and the number dropped from 52 mints striking type XIV to 23 for type XV. Similarly, the number of mints dropped from 45 striking type X to 23 for type XII. If type XII dies were first being issued late in 1124, not all the mints would have had dies by the time of the assize. Assigning the 1124 date to type XII is thus in agreement with four considerations: the solar rosette in the design, the last coin to be snicked before the assize, a drop in the number of mints after the assize, and the assignment of the last three types of Henry to a 9-year period (1126–35) rather than a single type to a 10-year period (1125–35).

The last three types of Henry raise an interesting consideration concerning their sequence. Type XIII features a central star in the reverse quadrilateral. Assuming the 1124 date for type XII, type XIII would have been issued in 1126 or 1127. Significant comets were seen in both years, although only the comet of July 1126 was recorded in Europe, and neither were recorded in English chronicles. The 1126 comet had a 4.5° tail. The 1127 comet was recorded in eastern chronicles as

Figure 396. Type XIII (1121) Figure 397. Type XIV (1123)

Figure 398. Type XV (1125)

stretching across the entire sky, and was visible between December 16, 1126, and January 14, 1127. The 1127 comet is more likely to have been depicted on the type XIII penny as the star, making 1127 a better date for its issue.

Two more significant celestial events took place during the reign of Henry. In October 1132 a comet with intense rays and a tail stretching for 45° was recorded in England, and on August 2, 1133, the path of a total solar eclipse crossed England and central Europe. William of Malmesbury associated the 1133 eclipse with the departure on that day of Henry I from England, who was never to return alive: "The elements manifested their sorrows at this great man's last departure. For the sun on that day, at the 6th hour, shrouded his glorious face, as the poets say, in hideous darkness, agitating the hearts of men by an eclipse."

Either or both these two celestial events were significant enough to have been depicted on Henry's coinage, and his type XIV penny has a star to the right of the king's bust and a central star on the reverse. The reverse star might represent the comet of 1132, and this would be consistent with the same motif found on type XIII, and the obverse star could represent the 1133 eclipse. Thus type XIV would have been issued in 1133. Alternatively, the stars on type XIV may have been immobilized from type XIII, and 1130 would be a better issue date.

However, if type XIV were issued in 1130, then type XV should have depicted either the comet or eclipse or symbols representing both events, unless it were issued prior to October 1132, when the comet appeared. The type XV penny depicts neither celestial event. One would expect the comet and eclipse to be depicted on Henry's coinage, and perhaps the next coin type was being planned, but was never issued because of the king's death in 1135. Two years after the eclipse seems too long a wait, however, to exploit the celestial events on coinage.

If these two events are indeed depicted on the type XIV penny, then one of two possibilities must be fact. Either the type XIV was struck in 1133 and type XV followed in sequence, or type XV was struck before type XIV, probably around 1130. In the former case, a six-year gap between 1127 and 1133 for type XIII is excessive, as the period of time was too long for normal changes in coinage design and the rarity of type XIII does not support an extraordinary length of production. In addition, the period between 1133 and 1135 is too short to support the number of coins known for types XIV and XV.

Thus, either types XIV and XV were struck prior to October 1132 and the two celestial events were not depicted on Henry's coinage, or type XV precedes type XIV. The evidence for the sequence of types XIV and XV as suggested by Brooke is strong. Although Brooke classifies the text style for the last two types as the same as the first two types of Stephen, opening the possibility of type XV preceding type XIV, a known type XIII/XIV mule suggests, but does not require, that type XIV immediately followed type XIII. In addition, but unknown to Brooke at the time, a type XV/Stephen type I mule exists. Brooke notes that three coin hoards were found containing Stephen pennies and only Henry type XV pennies, suggesting a link between the last type of Henry and the early types of Stephen. Some coin hoards contain coins of Stephen and earlier types of Henry, but none contain Stephen types and only Henry type XIV.

Blackburn (1991) asserts quite correctly, however, that the number of coins found in hoards does not fairly reflect the relative proportions in which different types were struck, and that the number of single finds of coins is a more indicative measure. Considering only single coin finds, Blackburn's research shows that type XV is by far the most common type of Henry's pennies, and that type XIV is one of the rarest. The relative rarity of type XIV could indicate that it was struck for a shorter period than originally thought, and perhaps after the 1132–33 celestial events and only until Henry's death in 1135. A limited period of production could devalue the type XIV/XV sequence conclusion derived from the hoard and mule evidence. Nevertheless, one must accept type XV as Henry's last type unless new hoards or other evidence are found to the contrary, and type XV is therefore dated 1132, prior to the celestial events.

Table 7.2
Revised Dates for Coins of Henry I

Type	Description	Dates	Precipitating event
I	Bust with annulets Cross fleury	1100–02	Death of William II
II	Profile left Cross fleury	1102–04	Unspecified
III	Bust with annulets PAX, four annulets	1104–05	Solar circles
IV	Facing bust Annulets, pyramids	1105–06	Unspecified
V	Bust with star Cross pattée, trefoils	1106–07	Comet, February 1106
VI	Bust with three stars Cross, saltire, stars	1107–10	Tinchebrai victory, September 1106
IX	Star, four annulets Cross, four annulets	1110–14	Planets of 1106, comet of 1110
VII	Facing bust Quatrefoil, pyramids	1114–17	Comet, May 1114
VIII	Bust left Annulets	1117–19	Unspecified
XI	Bust left, annulets Double inscription	1119–21	Unspecified
X	Facing bust Cross fleury	1121–24	Unspecified

7. Norman England

Type	Description	Dates	Precipitating event
XII	Bust with solar rosette Cross annulets	1124–27	Total eclipse, April 11, 1124
XIII	Bust left Star in quatrefoil	1127–30	Comet, December 1126
XIV	Facing bust, star Pellets in quatrefoil	1130–32	Unspecified
XV	Bust left Quadrilateral, cross	1132–35	Unspecified

The Coinage of Stephen and the Anarchy

During Stephen's reign and the anarchy caused by the civil war of 1138–53, four royal pennies and many irregular coins were struck. Some of the barons were given, or assumed, the right to strike coins in their own names; however, the moneyer's name and mint were not always present. Some moneyers took advantage of the unsettled conditions, struck coins of lower weight than required by law, and replaced the reverse legend with ornamental symbols, meaningless letters, and/or blundered strokes of partial letters so that the underweight coins could not be traced.

Other moneyers struck coins with blundered legends when they did not want to show loyalty to either cause for fear of choosing the losing side. In some instances, moneyers recalled as much of their own coinage as possible and hammered out their name on each coin. A few moneyers who did not have the ability or time to create or acquire new dies for their coins, defaced the bust of Stephen on existing dies to show their lack of support for the king. Some modern numismatists argue that coins with defaced dies were struck as a result of the papal interdict of 1148, but as Boon (1988) points out, defaced coins struck at Nottingham would have had to have been struck before the great fire there in 1140.

The first regular issue of Stephen has a bust with sceptre on the obverse and a cross moline on the reverse (figure 399). This type is now known as the Watford type, because of the large hoard of more than 600 coins of this type found in Watford, Hertfordshire, in 1818. The Watford type was issued in 1136 and continued through 1140–41, but may have continued to have been struck for several more years at mints that were cut off from royal centers where dies of newer types were made. During the civil war, Watford type coins were struck at Angevin controlled mints, but with the name of Empress Matilda or her son, Henry of Anjou.

Figure 399. Watford type

The greatest turmoil occurred during 1139–41, and barons struck their own coinage and supported whichever side of the royalist-Angevin war they thought was winning. Many of these local issues are similar to the Watford type in design.

Watford type coins were struck in northern England and Scotland as well. In 1136, David

I of Scotland moved south into England and captured Carlisle, thereby gaining access to the newly created English mint there and to locally mined silver. Some months later, David I established another mint in Edinburgh. Following the Battle of the Standard in 1138, the peace treaty signed in Durham in 1139 gave Prince Henry, son of David I of Scotland, the earldom of Northumberland. Several mints struck coins for Prince Henry, and at Bamburgh coins were first struck in the name of Stephen and then in the name of the prince. The early coins of David and Prince Henry, struck from 1136 to the early 1140s, are Watford type pennies.

The second regular issue of Stephen is the voided cross type with mullets in the four reverse quadrants (figure 400). This coin was struck only at the eastern and southeastern mints in England under Stephen's direct control, and may have been issued as late as 1145 (Blackburn 1994). However, it is more likely to have been issued shortly after Stephen's release from Bristol in November 1141 as maintained by some numismatists and supported by the mullets in the design.

Figure 400. Voided cross and mullets

The struggle for control of the English throne and support of the populace meant that both the royalists in Stephen's camp and the Angevins who supported the Empress Matilda sought every opportunity to validate their position. A show of strength and demonstration of a divine right to the throne were extremely important. Two solar eclipses, one on March 20, 1140 (figure 401), and the other on October 26, 1147 (figure 402), and Halley's comet of 1145, provided opportunities to claim heavenly support or to predict the demise of the king, and these celestial events were depicted on coinage struck by both sides during the civil war.

According to William of Malmesbury, the total eclipse of 1140 was remarkable:

> During this year, in Lent, on the 13th of the kalends of April, at the 9th hour of the 4th day of the week, there was an eclipse, throughout England, as I have heard. With us, indeed, and with all of our neighbors, the obscuration of the Sun also was so remarkable, that persons sitting at table, as it then happened almost every where, for it was Lent, at first feared that Chaos was come again: afterwards learning the cause, they went out and beheld the stars around the Sun. It was thought and said by many, not untruly, that the king [Stephen] would not continue a year in the government.

Many of the local and irregular Watford type pennies have stars, mullets, or rosettes in their designs that refer to one of the two eclipses. All of these coins are very rare, confirming that they were not struck as regular coinage by the king. One of the types has the obverse sceptre replaced by a flag and a mullet (figure 403). The mullet sometimes appears as a star, but presumably, the central hole of the mullet has been worn smooth on these pieces. This flag type coin was originally thought to represent the standard carried into the battle at Northallerton in 1138, suggesting an early date for the coin, but subsequent research (Boon 1988) asserts that the flag on the coin was copied from Stephen's second Great Seal, which he adopted some time during 1139–44, and thus the flag type coin was struck after 1138.

Some of the irregular coins of Stephen have either a rosette or a mullet in front of the king's bust or in the legend, suggesting that these two symbols were used interchangeably (figure 404).

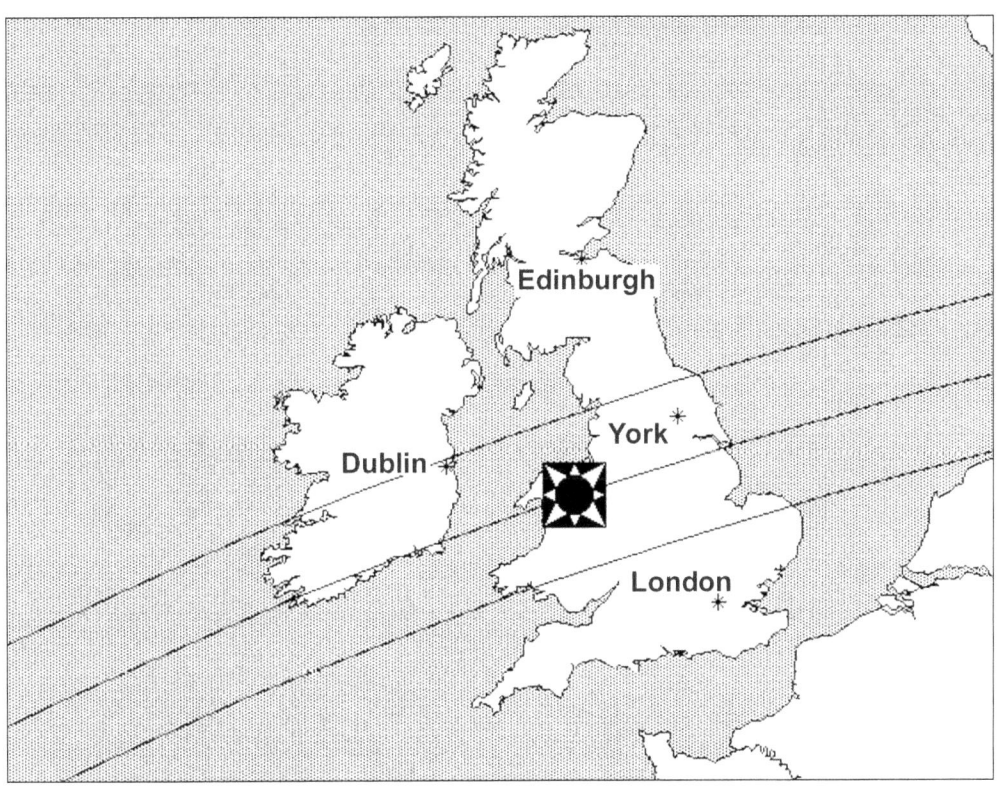

Figure 401. Path of the March 20, 1140, total eclipse

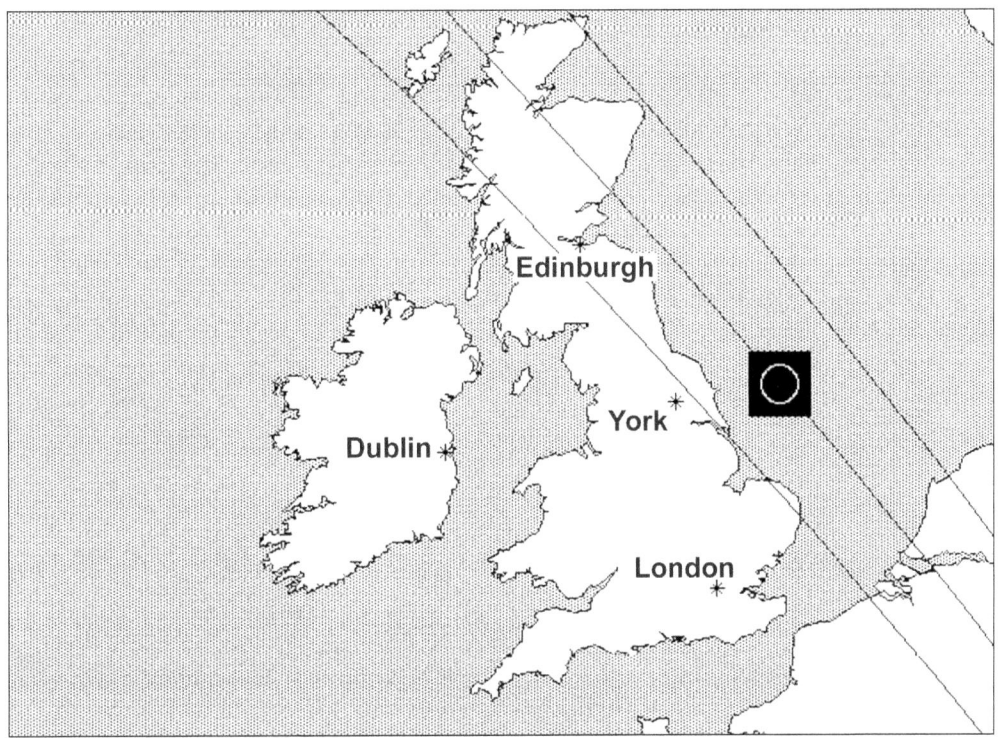

Figure 402. Path of the October 26, 1147, annular eclipse

Figure 403. Flag type

This supports the rosette as a solar eclipse symbol on the type XII penny of Henry I. Some of the rosette types have mullets in the reverse quadrants (figure 405), but with a different type of cross than on Stephen's second issue. Perhaps these pieces were either a transitional type between the first and second regular issues or local issues struck where second issue dies were unavailable. The mullet and sceptre type struck at Durham has annulets on the reverse fleurs and may be a reference to the 1147 annular eclipse (figure 406).

Figure 404. Rosettes in legend

Figure 405. Rosette and mullets

Figure 406. Mullet and sceptre

Figure 407. Henry of Blois

In a similar fashion, Henry of Blois, bishop of Winchester and the king's brother, struck a penny with a mullet in front of his crozier (figure 407). Mack (1966) suggests that the bishop's penny was struck between July and November 1141 while Stephen was being held captive and after Henry rejoined the royalist side of the conflict. This unusual penny, struck only four months after the total solar eclipse, combines the bishop's crozier with a celestial omen as a sign of support for the king.

At the Cambridge mint, Simon de Senlis issued coins (c.1141–45) in the name of Stephen, but with a large star at the end of the legend where a rosette is sometimes found (figure 408). Patrick of Salisbury, earl of Wessex (1141–47), struck a helmeted bust with sword type penny with a star behind the bust (figure 409).

Figure 408. Simon de Senlis

Figure 409. Patrick of Salisbury

An interesting penny has a mullet replacing the fleur at the top of the sceptre (figure 410). Mack states that some numismatic sources refer to this as a horseman's mace. Although a foot soldier's mace is depicted on the Bayeaux tapestry, the use of a horseman's mace during this period in unknown. A representation of the total eclipse is more likely.

Figure 410. Mulletted sceptre

Stephen may have wished to counter the public belief related by William of Malmesbury that the 1140 eclipse was the harbinger of doom for his reign, and to use the celestial event to his own advantage. The flag type and other local coinage with mullets and solar rosettes were most likely struck soon after the eclipse and perhaps as a show of support for Stephen while he was held captive. Coins with solar symbols that were struck after the 1147 eclipse served to reinforce Stephen's claim.

Although the voided cross and mullet type may have been struck after the eclipse and prior to Stephen's capture in February 1141, it is more likely that he struck his second coinage after regaining control of the throne in November of that year. The local coinage with solar symbols, especially the penny with the cross and mullets reverse (figure 405), would then be either transitional pieces struck before the second issue or local coinage struck during the early 1140s at mints that were cut off from receiving the second coinage dies.

Stephen was not the only king to make use of the 1140 eclipse to assert his divine right to a throne. The path of this eclipse crossed northern Germany, Denmark, and Russia. In Brunswick, Henry the Lion (1142–95) struck a silver coin with a mullet perhaps as early as 1140 (figure 411).

Not to be outdone by the royalists loyal to Stephen, several Angevin coins were struck that also depict mullets. One can assume that both sides in the civil war would have claimed divine right to the throne based on these celestial events. Clearly, a propaganda war was going on between the two sides in the conflict, and the depiction of celestial events on coinage was one of the media used for this purpose.

Figure 411. Henry the Lion

One type that has mullets on both sides of the bust is thought to have been struck by Brian fitz Count, son of the count of Brittany, lord of Wallingford, and lord of Abergavenny until 1141, when Matilda granted the latter title to another supporter. The coin is struck in the name of Brian, probably at Wallingford Castle (figure 412).

Figure 412. Brian fitz Count

Figure 413. Henry of Anjou

A similar coin was struck at Bristol in the name of Henry of Anjou as son of Matilda, because the title of *REX* is absent from the legend (figure 413). Boon suggests that the coin may have been struck in 1149 or 1153, and associates the mullets with the stars type of William the Conqueror. He also raises the possibility that the coin was the work of moneyers who were remaining neutral in the civil war by naming past kings on their coins. Henry was born in 1133 and the stars on this coin may be associated with the 1133 total eclipse or the eclipses of 1140 or 1147, thereby setting the stage for his divine right of accession to the throne.

Another coin was stuck at Wareham in the name of William (figure 414). If the coin were struck in the name of William de Mohun, earl of Dorset and Somerset, it would have been struck prior to 1144 when his support for the empress ended. This would be consistent with the total eclipse of 1140. Another possibility for the name in the legend would be that of William, earl of Gloucester, and the coin would have been struck after 1147 with the stars representing the eclipse of that year.

Figure 414. William

A Midlands variety has a reverse with a star within a quadrilateral and four annulets (figure 415). This coin is thought to have been struck around 1148, and most likely depicts the 1147 annular eclipse.

Figure 415. Midlands variety

The annular eclipse of 1147 may have been represented on several varieties of local coinage. A Watford variety of Stephen struck at Norwich has annulets with a central pellet in lieu of the fleurs (figure 416). An irregular Watford type with the bust facing left and blundered legends has the same motif replacing the fleur at the top of the sceptre (figure 417).

The Watford type coins with annulets may have been struck to counteract the message of divine right on coins of David I of Scotland, the Empress Matilda's uncle. The blundered legend penny, thought to be from northern England, may have been struck by a moneyer who feared to show allegiance to either king.

David's first coinage, struck between 1136 and the early 1140s is the standard Watford type, but with his name instead of Stephen's. Some time in the middle to later 1140s, he issued a coin with annulets enclosing pellets in the quadrants of the reverse cross (figure 418). These pennies may have been struck in reference to the 1147 eclipse.

Figure 416. Annulet reverse   Figure 417. Anuletted sceptre

Figure 418. Second coinage of David I

David struck one of his cross fleury type pennies with a rosette of annulets and a star in reverse quadrants (figure 419). The normal cross fleury type with a single pellet in each quadrant was issued in Scotland beginning in the late 1140s and through the end of his reign in 1153, and this star and rosette variety most likely represents the 1147 annular eclipse that crossed Scotland.

Figure 419. Star and rosette type

Now consider Stephen's third regular issue. This penny has a bust left with sceptre on the obverse and a cross fleury with pyramids with a trefoil of annulets (figure 420). As with Stephen's second coinage, his third type was struck only at eastern and southeastern mints. Most likely, other mints were cut off from receiving the new dies. North dates the issue of this type to around 1150, whereas Mack suggests a date as late as 1153, but certainly before the Treaty of Winchester in that year.

An irregular penny has a bust with sceptre facing right on the obverse and a voided cross

Figure 420. Third type of Stephen   Figure 421. Pyramid reverse

on the reverse with four pyramids joined in a quadrilateral (figure 421). The coin was found in a hoard that also contained the last two types of Henry I and Stephen's first type. The obverse suggests that it was struck during the same period as the Watford pennies, but the voided cross is more consistent with that of Stephen's second issue, and the pyramid is suggestive of Stephen's third regular coinage.

Six comets were recorded during the reign of Stephen, but only Halley's comet of April–July 1145 was recorded in European chronicles. The tail of the comet reached a maximum of 20°, and it seems unlikely that its appearance would not have been used to the benefit of both sides of the conflict. Perhaps this comet was depicted by the pyramids on either or both of these coins. If so, the third type of Stephen may have been struck as early as late 1145 or 1146.

Stephen's last coinage (figure 422) was struck after the Treaty of Winchester in 1153 and is named after the Awbridge hoard found in 1902. This type was struck at royal mints and formerly-held Angevin mints, and continued to be struck during the first four years of the reign of Henry II. Although it may have been struck first at royal mints before the treaty and then at all mints thereafter, there is no evidence to support an earlier date.

Figure 422. Awbridge type

Table 7.3
Revised Dates for Regular Coins of Stephen

Type	Issue date	Precipitating event
Watford	1136	Death of Henry I
Voided cross and mullets	1141–42	Total eclipse, 1140
Pyramid	1145–46	Halley's comet, 1145
Awbridge	1153	Treaty of Winchester

End of an Era

Thus ends the tumultuous period of Norman England, characterized by conquest, war, and heavenly omens. Celestial symbols were rarely used on English coins again, except on jetons (tokens), and as marks to denote new coinage, such as the star on the second coinage of Edward III that was introduced in 1335, perhaps in reference to the total solar eclipse of 1330 that crossed Scotland. However, in English controlled France and Ireland, and in Scotland, local customs prevailed and celestial symbols were used on coinage for another 200 years.

8

Re-attribution of Anglo-Gallic Deniers

Denier of Edward II

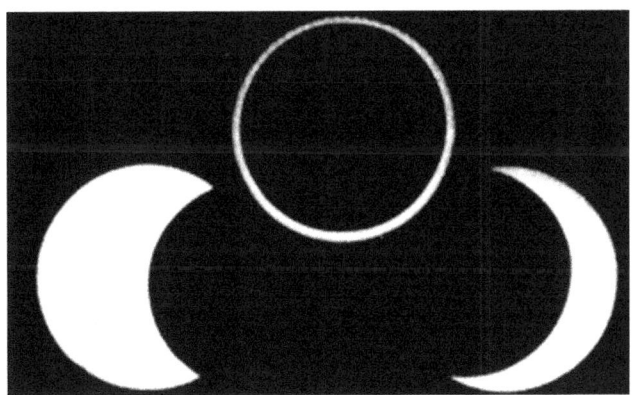

Annular solar eclipse

In the previous chapter on Norman coinage, the attribution and sequence of coins are well established, and celestial events depicted on coins were used to help date the series. In the regions of France controlled by England from 1152 through the end of the Hundred Years War in 1453, coinage at Anglo-Gallic mints was frequently interrupted for various reasons, and the economic history of this region remains somewhat unclear. To add to the confusion, from 1272 through 1377, three consecutive kings of England were named Edward (I, II, and III), in addition to Edward the Black Prince, son of Edward III, and none of the coins of this period distinguish between the three Edwards in the coin legends. Fortunately, the Black Prince is depicted on coins without a king's crown, and his title of prince is usually abbreviated in the legend, thereby making his coins distinguishable from his father's.

Quite a bit of research has been conducted during the past century on the Anglo-Gallic coins, and most, but not all, of the coins of Edward III and his son have been sorted out with some degree of certainty. Changes in style, coin weights, silver content, and hoard data have been used in the process. The early issues of Edward I and II struck in Aquitaine, however, and the status of operations at the Anglo-Gallic mints during their reigns, has been the subject of much debate. The standard reference work on the Anglo-Gallic coins was written by Elias (1984), and he attributes some issues with certainty, but only speculates on other coins. The discovery of a new Anglo-Gallic coin type by Ford in 2001 and subsequent joint research with this author that takes astronomical symbols on the coinage into account offers new insight into the sequence of issues and mint operations during the reigns of Edward I and II.

Transition from French Feudal Coinage

Henry, the future King Henry II of England, was the duke of Normandy and the count of Anjou, Maine, and Touraine. As a result of his marriage to Eleanor of Aquitaine in 1152, Henry received the additional titles of duke of Aquitaine and count of Poitou. Henry passed Aquitaine and Poitou to his teenage son Richard (the Lionheart) in 1172, and the balance of his French holdings went to Richard in 1189 when Henry died. Richard spent most of his boyhood in France, and lived in England for only five of his months as king. In 1195 Richard added another title when he captured Issoudun in the province of Berry. Upon Richard's death in 1199, his brother John inherited all these lands, but lost all except Aquitaine and part of Poitou to Philip II of France in 1202.

In 1252 Henry III of England installed his son Edward as duke of Aquitaine. Edward became King Edward I of England in 1272. The English also laid claim to the fief of Ponthieu when Edward's wife, Eleanor of Castille, inherited the county in 1279, and in 1283 Edward petitioned the French king for the right to issue coinage. This was granted, provided that the coinage was of similar stature to that of previous French feudal issues. Upon Eleanor's death in 1290, Edward II became count of the fief, but because the count of Alençon contested his claim to Ponthieu, the French administered the territory. Edward II issued coinage as count, and then as king until 1325, except for 1316–22, when the French again seized Ponthieu.

Both to ensure economic stability and as a requirement levied by the French kings, the Anglo-Gallic mints of Aquitaine, Poitou, Issoudun, and Ponthieu followed local customs when striking coins. This was partially due to respect for local traditions, but mostly due to the demand of the French kings to prevent the minting of high silver content coins such as those struck in England, thereby making royal and feudal French coinage undesirable by comparison. Other English-controlled French provinces such as Normandy did not strike coins in the name of the English monarchs, but continued to strike immobilized or anonymous local deniers. Many of the Anglo-Gallic coins were billon, which is a debased alloy of less than 50 percent silver. Consequently, the billon coins were rarely hoarded, and the mints often melted them down to provide metal for new issues. Thus many of the coins are very rare, making their attribution, sequencing, and dating even more difficult.

Astronomical symbols are found on some of the coins struck at all four Anglo-Gallic mints that were at one time or another under English control. However, the use of astronomical symbols on the coins of these regions began centuries prior to the Anglo-Gallic coinage.

Poitou

From the second half of the ninth century until the reign of Richard, Poitou's coinage was an immobilized design that was struck in the name of King Charles the Bald of France (figure 423). Some of these immobilized deniers have a mullet or crescent added to the design, but no correlation can be made with any specific celestial event. The symbols do confirm, however, that astronomical symbols were used on coinage of Poitou that preceded the coins Richard struck.

Figure 423. Immobilized denier of Poitou

Coins of Poitou struck by Richard (1189–99) are found with and without an annulet in one of the quadrants of the cross (figure 424). An annular eclipse crossed England on June 23, 1191 (figure 426), and was seen in Poitou as a thin crescent, but the annular eclipse of December 6, 1192 (figure 427), which crossed directly over Poitou, is more likely to be the event depicted on Richard's coinage. The varieties without the annulet were probably issued prior to the eclipse. Subsequent royal coinage struck by Alphonse de France (1241–71) after the confiscation of Poitou from John is identical in form to that of Richard, except that the annulet was replaced by a lis, suggesting that the annulet was not a symbol of Poitou, but rather a mark associated with Richard's reign (figure 425).

Figure 424. Richard   Figure 425. Alphonse de France

A contemporary of Richard I wrote:

> A wonder strange I write,
> The sunne did set, yet was no night.

Scott-Giles (1951) interpreted these lines as meaning that even though King Henry had died, the glory of the land was still bright, because Richard was another "sunne." These lines can

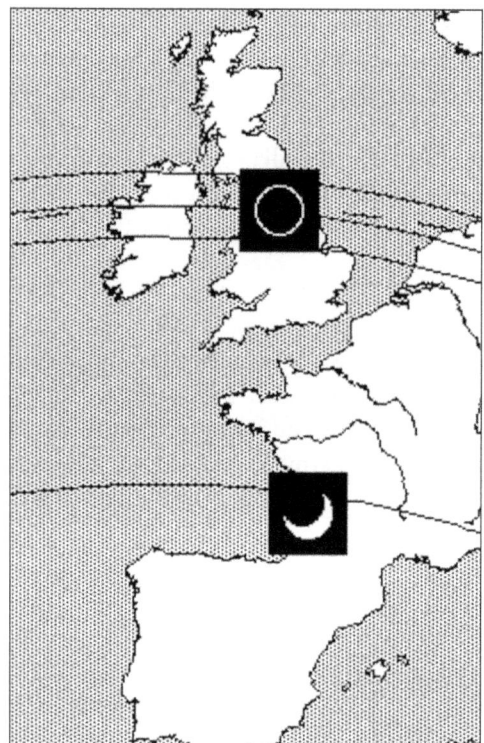 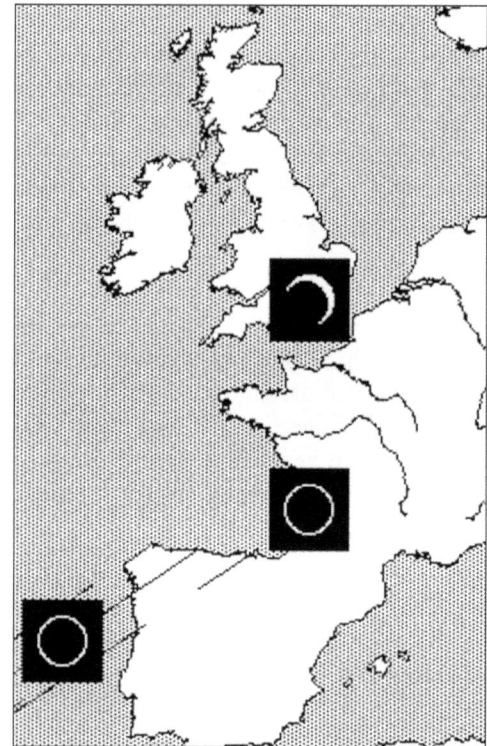

Figure 426. Annular solar eclipse, 1191   **Figure 427.** Annular solar eclipse, 1192

also be a description of the 1191 or 1192 eclipse, when the sun became dark even though it was daytime.

Numismatic rarity evidence by itself appears to contradict the assertion that the deniers with annulets were struck soon after the December 1192 eclipse. Poitou deniers without annulets are much more common than those with annulets, even though the eclipse occurred only 4 years into the 11-year period during which Richard's deniers were struck. Thus one would assume that the deniers with annulets would be the more common type. Although Richard struck coins in Poitou from 1189 to 1199, he was on crusade from 1190 until October 1192. Richard was captured by Leopold of Austria on his return trip, sold to the German emperor Henry VI, and held prisoner until February 1194, when he paid an enormous ransom in silver for his release. Perhaps no coin design changes were made until his return, or more likely, much of the available silver bullion and coinage struck after 1194 went toward his ransom payment and was subsequently melted down for use as German coinage.

John did not strike coins in Poitou after Richard's death; however, his vassal, Savary de Mauleon struck coins with his own name in Poitou, Angoumois, and Gascony. These deniers and the Irish coins of John bear crescents, probably representative of the 1207 eclipse that was seen as a thin crescent in these locations.

Issoudun

Richard's coins of Issoudun were struck after he captured it in 1195, and all his deniers contain annulets (figure 428). The annulets on these issues are consistent with the 1192 annular

8. Re-attribution of Anglo-Gallic Deniers 157

eclipse that crossed Issoudun. Richard's coins of Issoudun may be of immobilized design. As early as the 11th century, coins of Issoudun were issued with a bar, the letter *m*, and an annulet in the design. However, the issues of Eudes III (see figure 131), which immediately preceded those of Richard, replaced the annulet with a crescent, probably in reference to the great annular eclipse of 1153 that was seen as a partial eclipse there. Note that the issue of Richard retained a double crescent motif, but added the annulet, thereby maintaining the local custom as well as commemorating the 1192 annular eclipse.

Figure 428. Richard's denier of Issoudun

Ponthieu

Three Anglo-Gallic deniers were issued in Ponthieu. Elias theorized that one coin belonged to each of the Edwards, but celestial evidence suggests otherwise. The first one (type 24, figure 429) is attributed to Edward I, because it closely resembles the coinage of his French predecessor, Jean de Nesle, count of Ponthieu from 1251 to 1279 (figure 430). Both deniers have annulets in the design. Some of Jean de Nesle's coins have crescents instead of annulets. Annular eclipses crossed Europe in 1255, 1261, and 1263, and each of these would have been seen as a partial eclipse in Ponthieu. Although all three eclipses were recorded just across the Channel in England, none of them were recorded in French chronicles. On April 12, 1279, the path of an annular eclipse crossed England and 77 percent of the solar disk was covered as seen from Ponthieu. Edward may have taken this eclipse as an omen for his acquisition of Ponthieu that year, and in keeping with local tradition, included it in his design.

Figure 429. Type 24   Figure 430. Jean de Nesle

In England, Edward I introduced his new long cross coinage in 1280. Although his coins do not reflect the 1279 eclipse, the first English jetons (tokens), also struck in 1280 with dies that copied Edward's bust from the obverse of his pennies, have stars and crescents as the central theme on the reverse, with mullets (sometimes referred to as rosettes) by his bust and in lieu of legends (figure 431).

The second denier (type 26, figure 432), attributed to Edward II, includes the title of count in the legend, and therefore was probably issued between 1290 and 1307 before Edward II became king. This coin features crescents and annulets in the reverse field. This denier is exceedingly rare; indeed, no specimens are known to exist today. The coin was described and pictured

Figure 431. Early jetons of Edward I

in a set of books written by Poey d'Avant (1858–62), and all current references to type 26 are based on his text and figure.

Figure 432. Type 26

The design for type 26 was not new for Ponthieu. Gui I of Ponthieu (1053–1100) struck a denier (figure 433) containing annulets and crescents that is crudely similar to the type 26 denier of Edward II. The cross and *T* at each end form a shape similar to the elongated cross on type 26. On both coins, annulets and crescents are located above and below the cross, which may represent an annular solar eclipse. The paths of annular eclipses in 1064, 1084, and 1093 were close enough to Ponthieu that a thin, partial eclipse would have been seen there, and Gui's denier may depict any of these celestial events.

Figure 433. Gui I

The path of another annular eclipse passed directly over Ponthieu on September 5, 1290, the year Edward II became count, and the type 26 denier of Edward II depicts this event. Edward's claim to Ponthieu was being contested, and he may have issued this denier to assert his divine rights. This magnificent annular eclipse must have rivaled the great annular eclipse of 1153. The path of annularity ranged from 247 to 321 kilometers in width (figure 434). At its maximum view, annularity was visible for 6 minutes and 17 seconds. In Limoges, France, Godel (c.1320) wrote: "Also in this year [1290], on the day of Mars, before the Nativity of the Blessed Mary, in the first hour, with the moon at the 27th, there was an eclipse of the sun, with the sun itself being in Virgo, in the time of Pope Nicholas III."

Numerous examples of annulets on medieval coinage may depict the 1290 eclipse. Jean II

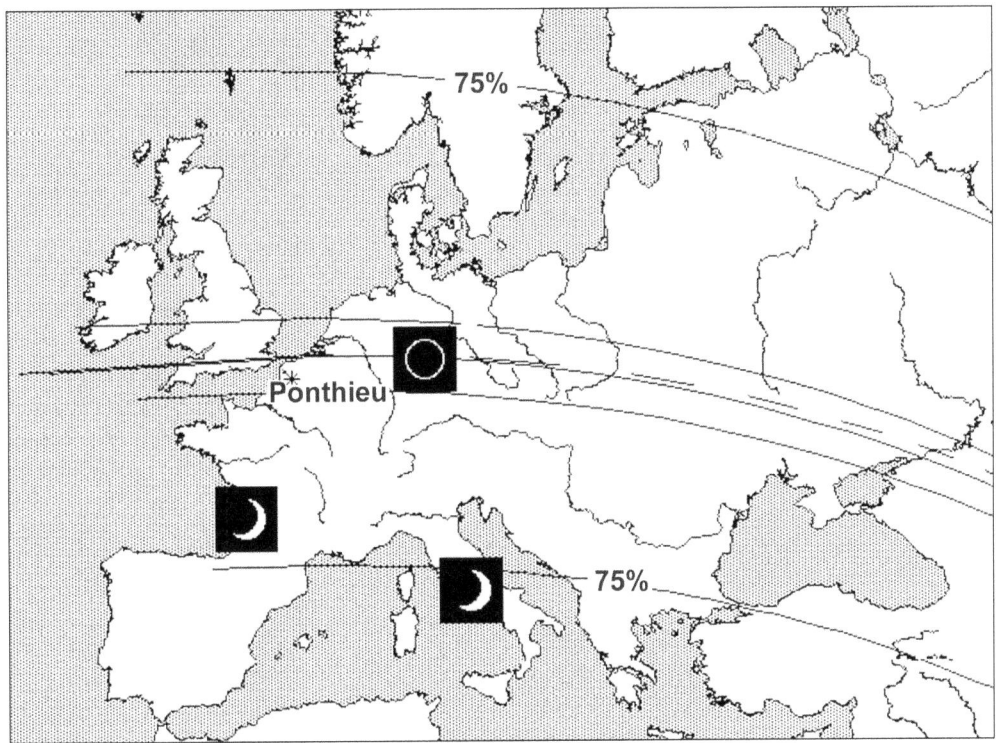

Figure 434. Path of the September 5, 1290, annular eclipse

of Brittany (1286–1305) issued deniers with a letter or an animal in the second quadrant of the reverse cross. One type replaced these symbols with an annulet (figure 435). The annular solar eclipse of 1290 crossed just north of the Brittany coast and was seen there as an 88 percent partial eclipse. To the south, in the French feudal province of Berry, Pierre I de Brosse (1287–1305) marked the early part of his reign with an obole containing an annulet (figure 436).

Figure 436. Pierre I de Brosse

Figure 435. Denier of Jean II

In England, Edward I's class 7 pennies (figure 437) issued in 1292 were modified to include a mullet or rosette on his breast, which suggests a reference to the 1290 eclipse. His class 5 penny was issued in 1289 and prior to the eclipse, and the class 6 penny is now thought to follow class 7 based on recent numismatic research. Note that Edward's class 4e penny, probably struck near the end of the 1282–89 period, has three pellets on his breast (figure 438). In 1278, 1280, and 1286, three planets came together in extremely tight conjunctions. The 1286 conjunction consisted of Venus, Jupiter, and Saturn, and on January 1, 1286, all three planets were contained in less than a 1° circle. This suggests that actual celestial events were represented on both pennies.

Figure 437. Class 7   Figure 438. Class 4e

Even though Elias attributed the third design (type 27, figure 429) to Edward III because of its similarity (leopard between two lines) to those of Aquitaine that he attributed to Edward III, he offered no explanation as to why no deniers of Ponthieu are attributed to Edward II during the 20 years he held the title of king. Varieties of the type 27 denier have annulets, crescents, or both in the reverse quadrants (figure 439). Hewlett (1920) also associated deniers of Ponthieu to similar designs of Aquitaine, but attributed this denier to Edward I.

Figure 439. Type 27 varieties

More likely the type 27 deniers were struck while Edward II was king. France reclaimed Ponthieu in 1337, and the only other annular eclipses crossing directly over royal or feudal France between 1290 and 1337 were in 1310 and 1321 (figures 440–441), both during the reign of Edward II. A partial eclipse may have been seen on November 8, 1295, but this event probably occurred too early to be represented on type 27. Interestingly, the eclipse of 1310 was seen as annular in Ponthieu and as a partial eclipse in Aquitaine, with the reverse being true for the eclipse of 1321. This would suggest that type 27 was issued after 1310. Perhaps the varieties with crescents were struck after the 1321 celestial event. Note that on coins of Aquitaine issued from 1325 to 1355, including those issued under Edward II, using an annuletted T in the legend was standard. Perhaps this lettering style was inspired by the annular eclipse of 1321, although it was used sporadically on royal French coins as early as 1298.

English jetons of Edward II struck between 1310 and 1337 are found with mullets, annulets, and crescents (figure 442). As these designs were not used on jetons of Edward I, they are more likely to represent the 1310 and 1321 eclipses than to be an immobilized design.

Figure 442. Jetons of Edward II

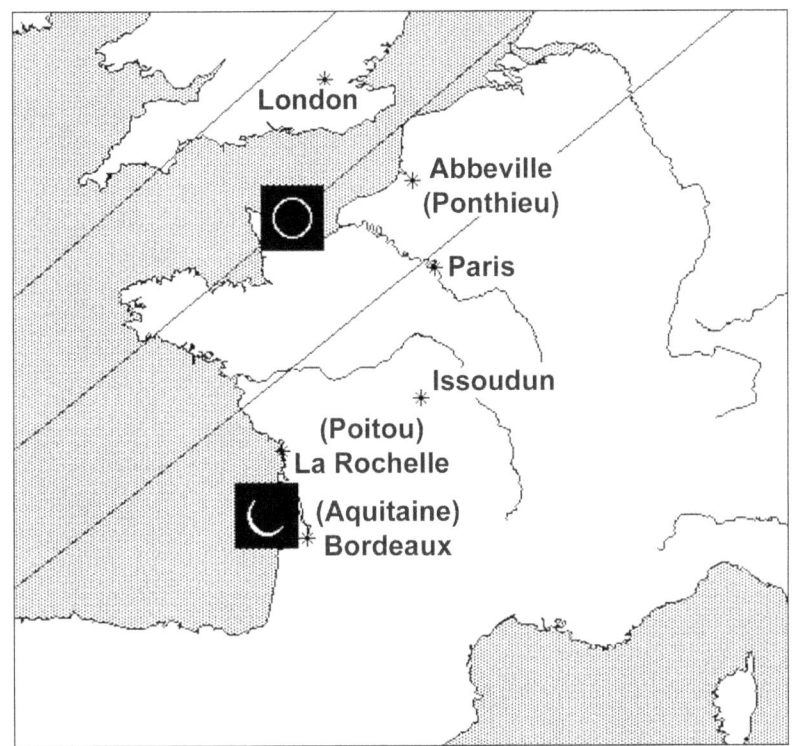

Figure 440. Path of the January 31, 1310, annular eclipse

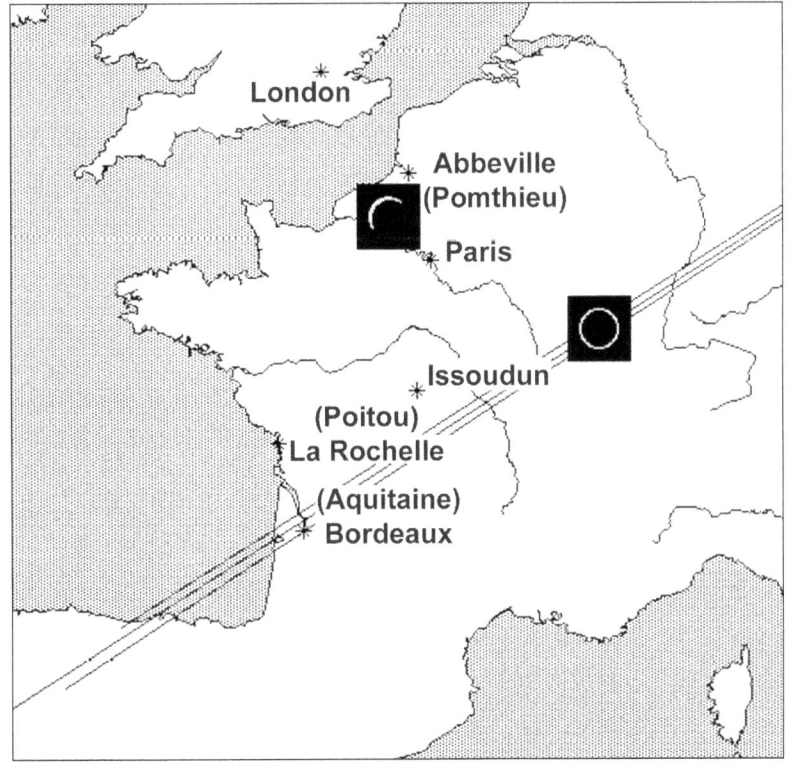

Figure 441. Path of the June 26, 1321, annular eclipse

Aquitaine

Chapter 4 has already discussed deniers struck by Henry II as duke of Aquitaine. All his coins contain annulets, and the motif can be related to the spectacular annular eclipse of 1153. During Richard's reign, coins of Aquitaine did not bear any royal title, suggesting that they were issued prior to 1189, and his coins do not depict any astronomical symbols. The latter is somewhat unusual, because a total solar eclipse crossed southern Aquitaine on September 13, 1178 (figure 443), and Richard's royal seals contained a star and a crescent.

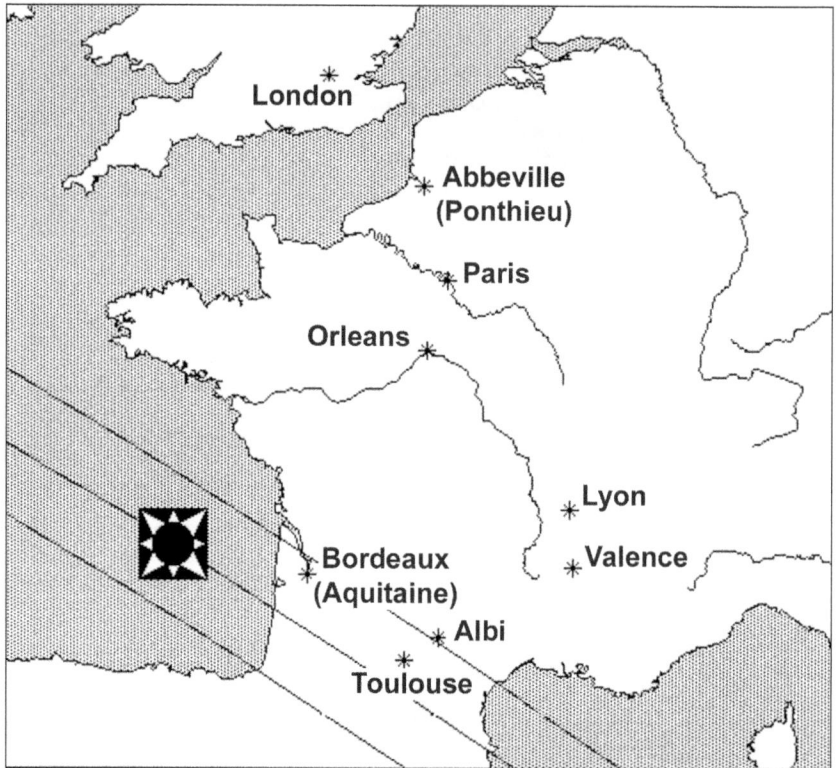

Figure 443. Path of the September 13, 1178, total solar eclipse

After paying homage to Louis VII for Aquitaine, Richard was installed as duke in 1172. In 1173 his mother, Eleanor of Aquitaine, encouraged Richard and his two brothers, Geoffrey and the English crown prince, Henry, along with a number of English and Norman noblemen and the kings of France and Scotland, to form a league against her husband, King Henry II of England. Henry II imprisoned Eleanor in 1174, leaving Richard to govern Aquitaine alone. Thus Richard's Aquitaine coins were probably issued after 1174 and before 1189. Perhaps the lack of any astronomical symbols indicates that his coins were issued prior to 1178. Alternatively, Warren (1973) wrote that Richard was proclaimed as the future lord of Aquitaine in 1170, but was not installed as duke until the summer of 1179, and therefore his deniers would have been struck after the eclipse.

Eleanor struck one denier after her formal restoration as duchess of Aquitaine in 1189. No Anglo-Gallic coins of John or Henry III are known, although one obole containing annulets that is currently attributed to Henry II may have been struck by Henry III based on other design considerations.

Re-attribution of the Aquitaine Coins of Edward I and Edward II

The previous discussion showed not only that the English-controlled mints in France issued coins that maintained local customs, but also that these coins often depicted actual celestial events. This practice was continued during the reigns of Edward I and Edward II, and thus may be used to help date the Anglo-Gallic coins.

Edward I struck deniers beginning in 1252 as duke of Aquitaine, and then after 1272 in his name as King Edward I of England. In 1285 Edward sent the English master of the mint, William de Turnemire, to Bordeaux to issue a new and better quality money for Aquitaine. This new money, the *moneta nova*, was delayed until 1291. Unfortunately, poor records, a closing of the mint for 11 years, and Edward II's accession to the throne all contributed to confusion about four known types of deniers and two types of oboles (half-deniers) that comprise the series of coins discussed here.

The *moneta nova* of 1291 came to an abrupt halt in 1294, when the French confiscated the duchy and occupied Bordeaux. In 1305, two years after Edward I's restoration in Aquitaine, the minting operations were resumed. Edward II became king following Edward I's death in 1307.

In 1984 Elias revamped many of the Anglo-Gallic coin attributions compiled by Hewlett (1920). However, this particular series of coins presented Elias with a number of unexplainable dilemmas, even though his ordering and attribution of the series used additional research and analysis to improve the conclusions. Hewlett doubted that any coinage was struck in Aquitaine during the reign of Edward II, but Elias referred to documentary evidence found by Rechenbach (1975) that clearly demonstrated that deniers, and perhaps oboles, were struck between 1307 and 1324. Nonetheless, Elias provisionally attributed this entire series of four deniers and two oboles to Edward I, but stated the possibility that some could have been continued by his son. The discovery of a fifth type of denier by Ford in 2001, combined with a plausible dating scenario based on celestial events and other factors, yields a new, simple, and logical ordering and attribution for the series.

The Coins

The first denier au leopard, type 18 (figure 444), depicts a leopard above the letters *AGL'* with an epsilon below and another epsilon in the first quadrant of the reverse cross. Elias could not explain the epsilon symbol, but nevertheless concluded that it is not part of the legend, because it is replaced by other symbols, such as a trefoil, on later deniers au leopard of Edward III (see figure 190).

The next denier, type 19 (figure 445), has the reverse epsilon in the second quadrant and is extremely rare: only two examples are known. The third denier, type 20 (figure 446), has the

Figure 444. Type 18

Figure 445. Type 19

Figure 446. Type 20   Figure 447. Type 23

epsilon returned to the first quadrant, with a crescent added in the second quadrant. Moreover, the legend below the leopard is changed to *AGI*. Type 20 is a fairly common variety.

Types 21 and 22 are oboles that have the same designs as denier types 18 and 20, respectively; however, a transitional subtype, 21a, reads *AGI'* instead of *AGL'*. Finally, a related issue (type 23, figure 447), is known as the denier à la couronne. This coin has a crown in lieu of the leopard, but like type 20 has *AGI* in the legend with an epsilon below, and like type 18 has an epsilon in the first quadrant of the reverse cross.

Elias attributed types 18 and 21 to the *moneta nova* of 1291 that was struck until the closing of the mint in 1294, with his dating of type 18 supported by hoard evidence that it was buried before 1305. Another hoard, buried after 1311, contained examples of type 20, and thus he attributed types 20 and 22 to the mint's reopening in 1305, and postulated that these types continued into the reign of Edward II. Because of the rarity of type 19, Elias speculated that this coin was struck just before the mint was closed in 1294. Type 23, also very rare, presented a dilemma for Elias. He attributed it to Edward I, but conceded that it might have been struck well into the reign of Edward II, or perhaps even between types 18 and 20, because it has the reverse of type 18 and the *AGI* of type 20.

The Fifth Denier, Type 20bis

The 2001 discovery of a new denier type, 20bis, simplifies the attribution sequence. This coin appears to be the same as type 20, but with the field of the reverse side of the coin rotated clockwise by 90° (figure 448). Likewise, the same rotation of type 18 would yield the type 19 reverse. In the case of type 19, Elias thought that the movement of the epsilon from the first to the second quarter was intentional and indicated a new and distinct issue. If his reasoning is followed for type 19, then type 20bis should be similarly interpreted. This interpretation, however, complicates the situation enormously by raising a number of perplexing questions about why and when these changes were made and how these various issues can be fitted into a logical sequence. If, however, the rotation of the field elements was unintentional, then the situation is greatly simplified. Consider the legends on these types. There is an initial cross at the beginning of the reverse legend and the letter *X* (of *DVX*) one quarter of the way into the legend. On types 18 and

Figure 448. Type 20bis

20 the epsilon is in the quadrant immediately to the left of the cross. On both types 19 and 20bis the epsilon is immediately to the left of the *X*. A die rotation mistake would have been easy for an engraver to make given the similarity between the cross and the letter *X*.

Types 19 and 20bis are exceedingly rare, which suggests that perhaps a single set of dies for each type was produced in error. The common denier types 18 and 20 both have corresponding oboles (types 21 and 22, respectively); however, no known oboles correspond to either type 19 or type 20bis. This could simply be due to the exceptional rarity of these two coins, or it could support the contention that these issues are indeed die rotation errors. In addition, the type 21a obole has the reverse of type 21 and the obverse of type 22, providing a direct link between the two. This would suggest that from a perspective of design development, there is no place in the sequence for either type 19 or type 20bis as intentional issues.

With the attribution of types 19 and 20bis as die errors, then only three known denier issues remain: types 18, 20, and 23, and there may have been a denier that corresponds to the obole type 21a. In addition, three dates are important: the *moneta nova* of 1291, the re-opening of the mint in 1305, and the accession of Edward II in 1307. These events do not, however, fully explain the three coin designs. If Elias were correct in his speculation that type 23 could logically fall between types 18 and 20 based on design considerations, then why would the crown replace the leopard when the mint was re-opened, and why would a crescent be added? Perhaps the answer can be found elsewhere.

Celestial Event Evidence

The epsilon and crescent on these denier types may represent one or more comets and a partial eclipse. An examination of the entire set of the series of denier au leopard types of Edward I through those of Edward III reveals that Roman *E*'s and Gothic epsilons appear on the same coin only on types 18, 19, 20, 20bis, 23, and on corresponding oboles. On all other issues in this series of deniers where the letter *E* appears on a coin, its form remains consistent. Yet on these particular coins the letter *E* always appears as a Roman letter in the legend, but as a Gothic epsilon in the field of the coin.

In this case, the differentiation seems to be intentional, and thus provides strong support for the contention that the epsilon in the field is a symbol, perhaps for a comet, rather than a letter. Significant comets were taken as omens for kingdoms changing hands. On coins of similar design, such as the denier of Raymond III of Orange (1335–40), a mullet is found in lieu of the epsilon, which suggests that this location in the field might have been used for an astronomical symbol (see figure 189).

During the reign of Edward I in Aquitaine, 10 comets were recorded between 1264 and 1277, 5 between 1293 and 1299, and 9 between 1301 and 1307. Of these, comets seen in 1264, 1265, 1269, 1301 (Halley's), and 1305 were both visually impressive and were recorded in Europe. Comets seen in 1273, 1277, 1293, 1297, 1299, and 1304 were also impressive, but were recorded only in eastern chronicles. The two comets seen in 1299 had long tails, one measuring 30° in length. In England, Edward's class 9 penny has a star on the king's breast (figure 449). This penny is thought to have been issued in 1299, and may depict one of the comets seen that year, or perhaps the penny was issued later and depicts Halley's comet of 1301.

The mint in Aquitaine was closed during 1294 and 1305, so the epsilon on these deniers cannot correspond to the star on the class 9 penny. However, this penny does indicate that Edward may have depicted comets on his coinage. Note that stars (and pyramids) were used on Norman coinage to depict comets, but that stars were rarely used for comets on the continent, but rather

Figure 449. Class 9 penny of Edward I

as solar symbols. In feudal France pyramids, epsilons, and combs were the symbol of choice for comets, and therefore Edward's use of an epsilon in Aquitaine would be in line with the local customs. Of course, the issue here is that the *moneta nova* was issued in 1291, and the comet of 1293 would have been visible too late to be incorporated into the design, while that of 1277 was too early, unless the *moneta nova* coinage had been designed several years prior to William de Turnemire's departure for Aquitaine in 1285. Also, the passage of 16 years without a comet sighting is unlikely, but this cannot be verified.

Crescents, annulets, and stars on English jetons of Edward II (figure 442) and on deniers of Ponthieu (figure 439) representing the eclipses of 1310 and 1321 give strong reason to believe that the crescent found on both types 20 and 20bis represents an eclipse as seen from Aquitaine.

Dating the Series

One can begin by accepting Elias's arguments for dating type 18 as the *moneta nova* of 1291. We can also accept his arguments that hoard data places type 20 after 1305. The annular eclipse on January 31, 1310, that crossed the northern coast of France, including areas just to the north and east of Paris and the English-held region of Ponthieu, was seen in Aquitaine as a thin partial eclipse. The 1310 eclipse was the only partial eclipse seen in Aquitaine during the reigns of either Edward I or Edward II after 1305. Therefore type 20 must have been minted after 1310, and is attributed to Edward II. Aquitaine was in the path of annularity of another eclipse in 1321, but this eclipse occurred too late for the type 20 denier. This is clearly the case inasmuch as the design symbol would have been an annulet rather than a crescent, and no evidence supports dating this coin after 1321.

Finally, one must address type 23. Here the lack of a crescent on the reverse would date it prior to the 1310 eclipse. The obverse design change could be indicative of the coronation of Edward II, hence placing it in the period 1307–10. This implies that the following sequence is plausible:

1. Both type 18 and the error type 19 are now assigned to Edward I in the period 1291–1307.
2. Type 23 is an early issue of Edward II dated 1307–10 that was not accepted by the people.
3. Type 20 and the error type 20bis are of Edward II and dated 1310–21.

This seems to be a coherent and logical resolution to the problems associated with these issues. Moreover, with type 19 disregarded as an error, type 23 now seems to be a natural fit between types 18 and 20. It has the reverse of type 18 and the obverse of type 20 with the lion replaced with a crown, perhaps to signify Edward II's coronation. One can then assume that for some reason type 23 failed to gain public acceptance and was replaced with type 20 in a reversion to the denier au lion. Here the addition of a crescent not only marked a new and distinct issue, but also symbolized the 1310 eclipse and reassured the public of Edward II's divine right to rule.

Design Development Versus Weights and Alloys

The overall appearance of type 23, particularly in relation to weight and alloy, indicates that it was debased. One could argue that this indicates that type 23 postdates type 20 because types 18 and 20 were of similar weight and fineness. Consider the possible coin sequences. Hoard evidence places type 18 as the *moneta nova* of 1291 and type 20 in circulation after 1311. Thus, two plausible sequences are possible, namely:

1. types 18/19 → 21a → 23 → 20/20bis, or
2. types 18/19 → 21a → 20/20bis → 23.

From a perspective of design development, the first sequence is intuitively more reasonable. The progression is natural. The second sequence, however, raises a disturbing question. The crown marks the difference between type 23 and earlier issues, so there would be no need to drop the design element of the crescent. Thus from a design perspective, the most simple explanation favors the first sequence.

Now consider weights and alloys. One should assume that whenever weights or alloys were intentionally changed, some change should also have been made to the design or privy mark to allow mint and government officials to differentiate between the old and new issues. Known specimens of types 18 and 20 are of comparable weight and composition. Type 23, however, appears to be debased. Moreover, the heaviest known specimen of type 23 is lighter than most known specimens of either types 18 or 20, thus types 18 and 20 were probably issued at a higher metal standard than type 23. Thus for both the first sequence and the alloy and weight analysis to be correct, one must conclude that a debasement in currency was followed by a return to the earlier standard. Research by Rechenbach (1975) resolved this issue.

Rechenbach's Research

Elias indicates that the exchange rates of the 1291, 1305, and 1322–23 issues were constant at five denier per English sterling penny. In her doctoral dissertation, Rechenbach confirmed these rates. Furthermore, North (1994) states that the value of the sterling during that period did not change. Thus these three issues should be equivalent. That should mean one of the following: (1) the apparent debasement of type 23 is incorrect, (2) type 23 was debased and later coinage was restored to the original standard, or (3) type 23 was struck after 1324 when coinage was reported as debased.

Rechenbach presents further details about the issues of Edward II based on mint records. She sets forth the following mint activity chronology, but does not assign specific coins:

1. Earliest issue during 1309–11
2. Continued mint activity during 1311–15
3. Mint idle during 1316–19
4. Richard of Ellesfield's coinage of 1319
5. Adam de Lymbergh's issue of 1322–24
6. Debased issue as a result of the St. Sardos war in 1324
7. Lapin Roger's tower issue of 1325
8. Bordeaux revision of the tower issue first struck in 1326.

Elias identified coins struck during periods 5, 7, and 8 as other coin types struck after the type 18–23 series. Although type 23 might be associated with the Ellesfield or post–St. Sardos coinage, the period between 1309 and 1311 provides the most plausible dates, as it can account for both design development and the historical record.

In addition to the chronology of mint operations, Rechenbach makes two other relevant observations from original records. First, she states that the records show that in 1311 Bernard Mandavini, the master moneyer at Bordeaux, was charged with debasing the coinage during the early issues of Edward II. Subsequent measurements demonstrated that the charge was unfounded. This would, nonetheless, support the contention that the apparently lighter weight type 23 denier was the earliest issue of Edward II, and that it was replaced with type 20 to restore public confidence in the coinage. Second, she points out that the records also show that Edward II brought in Italian moneyers to operate the mint in Bordeaux. Although circumstantial as evidence, note that Italian moneyers introduced the crown as a monetary design element in other European countries at the beginning of the 14th century.

Contemporary French Coinage

Different kingdoms in close geographic proximity often used similar designs for their coinage. The consistency of design helped assure the general populace of acceptance of the coinage as a medium of exchange. In January 1311, Philippe IV of France issued three new, debased coins. All three coins are particularly significant to the types 18–23 analysis. The bourgeois fort (figure 450) prominently shows a large crown similar to that on type 23. This is additional support for dating type 23 early in the reign of Edward II. The smaller bourgeois simple (figure 451) added a new symbol to the coins of Philippe IV: two annulets. Furthermore, some of the oboles bourgeoise (figure 452) contain an annulet above the obverse cross. Royal French lands were in the path of annularity during the 1310 eclipse.

Figure 450. Bourgeois fort   Figure 451. Bourgeois simple

Figure 452. Obole bourgeoise

In Poitou, no coinage was struck after 1271 when it was reunited with royal France. In 1311 Poitou was given to Philippe (le Long) de France, who ruled there until 1316 when it was again reunited with France. Poitou was just to the south of the path of annularity of the 1310 eclipse, and annulets flank each side of the lis above the chatel on his denier (figure 453). Just to the north of Poitou in the province of Maine, Charles de Valois (1290–1317) struck a denier with a crown on the obverse and annulets on the reverse (figure 454). The annulets may represent the 1290 annular eclipse, but the 1310 eclipse crossed directly over Maine and is the more likely representation. Thus the similarity between coins that are closely related in both issue date and geographic location provides additional evidence to support the crescent on type 20 as representing the eclipse of 1310 and type 23 with the crown being struck at about the same time.

Figure 453. Philippe de France   Figure 454. Charles de Valois

Conclusion

All the available evidence taken together — the new denier type 20bis, design development, weights and alloys, celestial events, comparison to royal and feudal French coinage, and documentary evidence of mint operations — indicates that the most consistent and plausible dating sequence and chronology for these issues is as follows:

Edward I: types 18/19 (1291–94), varieties of type 18 and/or 21a (1305–07),
Edward II: type 23 (1309–11), types 20/20bis (1311–15).

The Solar Crusade

Gros tournois with star of Louis IX (1266) issued to pay the armies of the Eighth Crusade

9. The Solar Crusade   171

The mysticism attached to celestial events was used not only to sway the minds of the general population, but also the decisions made by sovereigns. This chapter will show how such an event not only convinced Louis IX of France to go on the Eighth Crusade, but also convinced him where to land his forces. Only a few facts are known about what factors lay behind his decision making process, but Louis's coinage gives us clues that may provide some of the answers.

Historical Background

In 800 Pope Leo II created the Holy Roman Empire when he conferred an imperial crown on the French King Charlemagne. This linked Italy to the emerging states of northern Europe rather than to the Byzantine Empire. By the middle of the 11th century, northern Italy belonged to the Holy Roman Empire, central Italy was under papal control, and Byzantines controlled the south. By 1130, the Normans had grown strong enough to take control of Sicily, and then later moved into Naples. A century later the pope offered the two Sicilies, Sicily and Naples, to Charles, count of Anjou and the brother of Louis IX of France. Charles occupied the territory in 1266 after defeating Manfred, and in 1268 conquered the rightful heir, Conradin, and claimed the throne. The total solar eclipse of May 25, 1267, that crossed Sicily may have been taken as an omen for Charles's victory and as a sign of his divine right to the throne. This eclipse crossed northern Africa just to the south of the islands in the western Mediterranean sea and passed directly over Sicily, Greece, northern Turkey, and other parts of Asia (figure 455). It was recorded in Germany, Austria, Switzerland, and Byzantium (Constantinople). The Sicilian reale d'oro of Charles of Anjou, now King Charles I of Sicily, contains a star by his bust (figure 456), but the same coin struck by him in Barletta, located in northern Apulia, does not show the star.

Figure 456. Reale d'oro of Charles I

After his coinage reform of 1278, Charles's new carlino d'oro contained mullets, stars, and a crescent (figure 457). The carlino was minted in Naples, and the crescent probably refers to the appearance of the eclipse from there.

Charles was not the only king in the region to represent this eclipse on coinage. The kings of Aragon held various territories outside Catalonia. To the west of Sicily, on the island of Majorca that they had acquired from the Moors, James II (1276–1311) issued the first Aragonese coinage for the island. All his coins contained stars and/or mullets. His reale d'oro has a star to the left of the enthroned king and mullets on the reverse (figure 458). James II was to become king of Sicily in 1286, but relinquished the island for control of Sardinia and Corsica in 1297. All his Sardinian coins contain mullets (figure 459). The mullets on the coins of these Mediterranean territories most likely represent the 1267 total solar eclipse that crossed just to the south of the islands.

172   Astronomical Symbols on Ancient and Medieval Coins

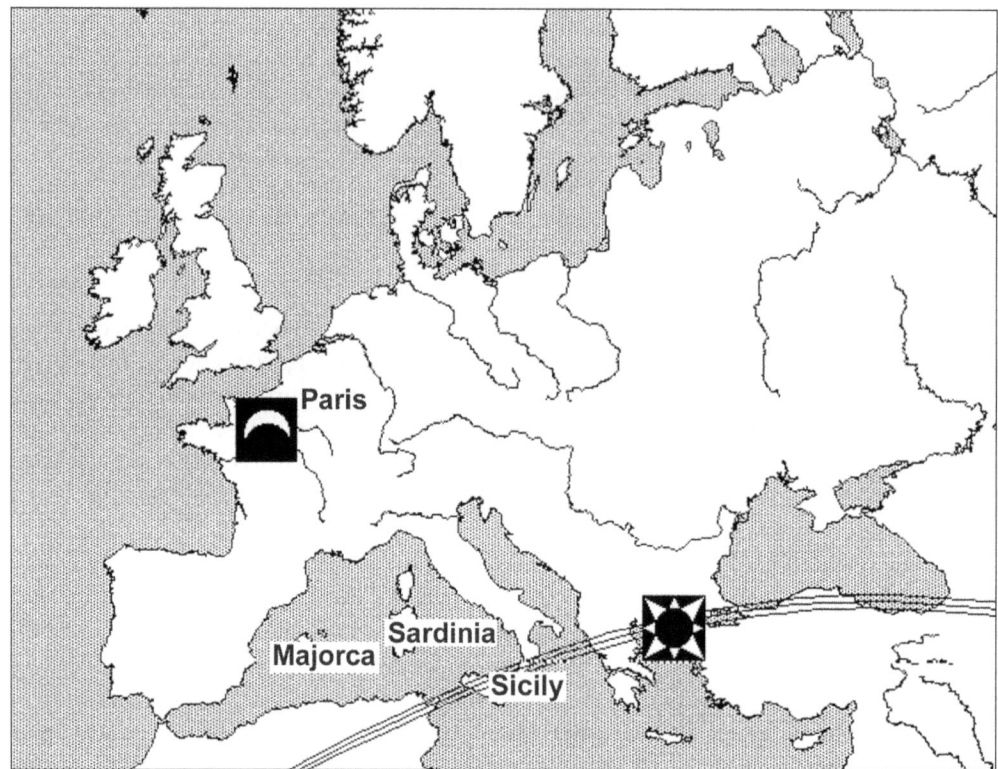

Figure 455. Path of the May 25, 1267, total solar eclipse

Figure 457. Carlino d'oro of Charles I

Figure 458. Reale d'oro of Majorca   Figure 459. Sardinia

In Byzantium, some of the coins Michael VIII (1261–82) struck have a star to the right of the enthroned emperor (figure 460). Constantinople had been recaptured by the Byzantine aristocracy in 1261, and Michael regarded Charles of Anjou as his bitterest enemy (Sear 1987). Although an astute diplomat, Michael needed to marshal all the forces he could to counter the threat from the Sicilian king. As soon as Charles used a symbol of the 1267 eclipse on his coinage, Michael would have countered with similar symbolism on his own coins to proclaim his divine rights, and because the eclipse had also been witnessed in Constantinople.

Figure 460. Michael VIII

The Crusades and Celestial Omens

The association of astronomical events with omens for the crusades began with the First Crusade (Billings 1990). Following the Council of Clermont, Pope Urban II launched the idea for the First Crusade on November 27, 1095. Seven months prior to the council, a meteor shower was seen throughout France. On February 11, 1096, the French king and his nobles, who were meeting to makes plans for the crusade, witnessed a spectacular lunar eclipse. The next month an aurora lit up the skies and resulted in the recitation of special prayers, and in August 1096 another lunar eclipse took place. All these heavenly events were taken as favorable omens for the crusade. As Baldwin's army marched into Syria in October 1097, his soldiers saw a comet with a sword-shaped tail, further reinforcing their belief that God supported their mission.

Louis IX may also have interpreted astronomical events as omens for his Seventh Crusade (1248–54). Duplessy (1999) cites the coincidence that a devalued French denier tournois (dies made at Tours) was apparently issued for the first time between 1240 and 1250 as Louis IX prepared for his expedition during 1245–48. Perhaps this coin was designed to pay for the crusade, and the devaluation was necessary to cover the great expenses of the preparations. Analysis of astronomical symbols on this new denier tournois requires consideration of the denier tournois that it replaced.

The first denier tournois (figure 461) was issued during the short reign of Louis VIII (1223–26) and the first part of Louis IX's reign (1226–45). This denier differs from the subsequent denier tournois (figure 462) Louis IX issued because it does not have an S at the end of the reverse legend, "TVRONVS CIVI." Versions of the first issue use either an epsilon or an *E* in the title "REX." The new denier tournois has only an *E*, suggesting that the *E* type deniers of Louis VIII and Louis IX may have been struck late in the 1223–45 period and only during the reign of Louis IX. All these types have varieties with a pellet on the *N* in the reverse legend.

Some varieties of the new denier have annulets in the legend. A rare variety of the denier of Louis VIII and IX (figure 463) has an annulet after his title "REX" in the legend of the coin and on the *N* of "TVRONVS." This rare type may have been struck shortly before the issuance of the new denier, as none of the other early types have the annulets. After the annular eclipse

174 Astronomical Symbols on Ancient and Medieval Coins

Figure 461. Denier of Louis VIII and Louis IX   Figure 462. New denier tournois of Louis IX

of 1207, the next one to cross France was in 1270, and the new denier tournois of Louis IX was issued between 1245 and 1250, thereby discounting the annulet as a solar eclipse symbol. Although partial solar eclipses may have been seen in France each year from 1239 through 1243, crescents would have been the more appropriate symbol in the design. So why would some of the new deniers, as well as the rare, early type denier of Louis VIII and IX contain annulets?

Figure 463. Annulet at end of legend

Perhaps the annulets represent a different type of astronomical event taken as an omen for the Seventh Crusade. The year 1244 was significant for Louis IX. In December of that year he took the cross and pledged to lead his first crusade. On February 25 and August 19, 1244, spectacular total lunar eclipses were visible in France, with each eclipse having a totality phase in excess of 100 minutes, during which light from the sun would have been refracted by the Earth's atmosphere, turning the full moon blood red in color. A hundred and forty-eight years earlier, two total lunar eclipses in the same year, also in the months of February and August, were seen as giving divine blessing to the First Crusade. Thus the annulets found on the rare variety, early denier and on the new denier of Louis IX may represent the total lunar eclipses of 1244. The annulet was also used in the crusader coinage of Louis IX struck in Palestine in 1251 and 1253 (figure 464). The Christian legends were written in Arabic. The design depicts a cross within a circle (annulet) within a square.

The Seventh Crusade (1248–52) turned out to be disastrous for Louis IX. He was soundly defeated in Egypt, became ill with malaria, was captured, and paid a heavy ransom for his release. Louis IX had always intended to undertake another crusade after this failure, but needed the proper incentive.

Figure 464. Crusader coinage of Louis IX

The Solar (Eighth) Crusade

Louis's brother, Charles of Anjou, was an aggressive conqueror. While Louis was a deeply pious ruler, Charles constantly sought new sources of power, and believed that he was God's chosen instrument (Runciman 1958). Furthermore, it may have been Charles who persuaded his brother to undertake the Eighth Crusade. The Moors held all of North Africa. Louis IX was reported to have received letters that made him think that he might be able to convert the dey of Tunis to Christianity. Were the letters from his ambitious brother, Charles of Sicily?

For years the goal of the kings of Sicily had been to create an empire in the lands bordering the eastern Mediterranean. Charles adopted this policy from Manfred, but was prevented from going after Constantinople until he had dealt with Conradin's claim to southern Italy. The Byzantine Emperor Michael VIII, concerned about an invasion by Charles, appealed to the pope in 1267 and pledged his support for a future crusade. The pope promoted this new crusade to the powers of Europe, and in 1267, Louis IX agreed, but found little enthusiasm among those around him.

After defeating Conradin in 1268, Charles began his preparations to invade Constantinople in 1270. Fearing the worst, Michael VIII sent messengers to Louis IX in 1269 suggesting a union between the churches of Rome and Constantinople and re-energizing Louis's thoughts about launching a crusade.

Louis sought the help of his brother, Charles, for the crusade. Now Charles had to choose between helping Louis and attacking Constantinople. He chose an alternative that benefited him. Tunis was just across the sea from Sicily. From the time of Roger II (1105–30) the rulers of Tunis had paid the kings of Sicily a yearly tribute in gold. Mustansir, the king of Tunis, discontinued the payment when Manfred was killed, citing the dynastic change in Sicily and claiming that Charles was therefore not entitled to the payments. In addition, Mustansir had given refuge to exiled supporters of Manfred and Conradin and was supporting rebellion against Charles in Sicily.

Charles directed Louis's attention to Tunis. Runciman states that Charles pointed out to Louis that Tunis would be valuable for an attack on Egypt and eastern Muslim-controlled regions. He also told Louis that the Tunisian king might convert to Christianity, but feared reprisals from his generals and clerics. Whether or not Charles believed this, Louis did, although his French advisors were less enthusiastic. They suggested to Louis that he go east where Christian reinforcements were needed. Charles would have preferred to go after Constantinople, but control of the Mediterranean shipping routes was of military value to him, so he needed to persuade his brother either to forego the crusade or to go to Tunis.

Louis decided on Tunis. He launched the Eighth Crusade from France in 1270 and landed at the Bay of Tunis on July 18. Mustansir made no effort to covert to Christianity, and while Louis and his army camped on the ancient site of Carthage, dysentery and typhoid swept through his troops, and on August 25, illness claimed the king of France. His brother Charles and Edward of England arrived in Tunis with their armies three days after Louis IX's death.

Perhaps the total solar eclipse of 1267 was also an omen for Louis IX, one that had convinced him that the time was ripe to try again. Charles would have known that the total solar eclipse of 1267, his omen for victory in Sicily, had also crossed directly over the Bay of Tunis and Carthage (figure 465). Perhaps he convinced Louis that victory at Tunis had been predetermined by the heavens.

The total eclipse of 1267 may be the basis for a type of gros tournois (four deniers) of Louis IX of France. His gros tournois was first issued around 1266 and continued until his death in 1270. One variety has a star between the legend and border below the châtel on the reverse of the coin (figure 466). The ordinance creating the gros deniers tournois d'argent is unknown, but it was

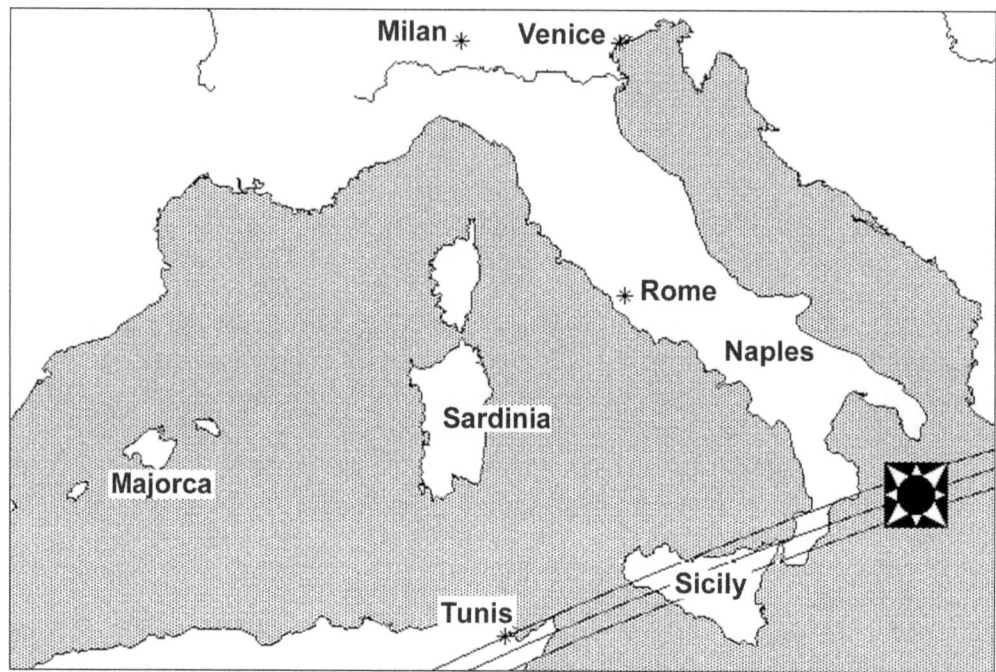

Figure 465. Path of the May 25, 1267, eclipse over Tunis

first cited in a document dated July 24, 1266, and was probably created along with the ecu d'or, although the ecu d'or was unknown as late as September 1268.

Figure 466. Gros tournois à l'étoile of Louis IX

These coins were probably issued to pay for the preparations for Louis IX's Eighth Crusade. The only known contemporary text that documented the ecu d'or was Sicilian. The coin was recorded in a list of coins found in the port of Trapani, Sicily, after a storm had destroyed part of the navy that returned from the crusade. If the large silver and gold coins had been issued to support this crusade, then perhaps the gros tournois with the star was struck for payment to Louis's army of 10,000 men and the ecu d'or was sent to Charles for his support.

The star could have been used to represent the 1267 eclipse as a propaganda device to build military morale and help convince the troops of divine intervention. A corresponding silver gros struck by Charles has mullets in the legends and in lieu of one of the annulets under the triangle of the châtel (figure 467). The innermost obverse legend reads "KAROLVS*SCL.REX" (Charles, king of Sicily). The reverse legend is "COMES*PVINCIE" (count of Provence).

Boudeau (c. 1920) states that this coin was issued in mid–1267 for use in Avignon when Charles became the master of Sicily. Clearly, the mullets represent the 1267 eclipse.

Figure 467. Gros tournois of Charles of Anjou

Louis IX chose to go to Tunis on his final crusade because of his brother's political maneuvering and encouragement and a promise of divine intervention in the form of a solar eclipse. This hypothesis is supported only by the limited and circumstantial evidence presented in this chapter. We may never know the contents of the letters that Louis received, but Louis IX and Charles of Anjou clearly placed a solar symbol on their coinage. Throughout, this book has shown that it was common medieval practice for sovereigns to garner political support based on celestial events and to commemorate those events on their coinage.

Beginnings

Denier of Sens (mid–11th century)

Viewing Halley's comet, 1066

10. Beginnings

I had originally planned to entitle this chapter "Conclusions," but that would have implied an ending. However, summarizing the findings presented in this volume as the first steps toward a new understanding of the political and mystical mind-set of ancient and medieval people is more appropriate.

The use of astronomical symbols on coinage, especially on the low or common denominations coins, ensured that the populace would constantly be reminded of the divine authority of the current sovereign. In an era of political turmoil in Europe, especially when kingdoms changed hands, rulers had to gain the support of a population that was illiterate, but that would recognize the symbols representing divine events that they themselves would have witnessed with wondrous amazement.

No doubt some of the examples presented in the preceding chapters are incorrect, whether because of erroneous attribution to a specific event or failure to recognize some form of immobilization or design copying. However, many of the examples are correctly attributed to physical astronomical events. With additional research, many more correct interpretations can be discovered.

Coinage design of each issue must have been conceived in the context of the thinking and political climate of the times, with some allowances made for artistic license. That design license could not have strayed too far from reality, or from desired political objectives. Failure to capitalize on the mystical beliefs of the masses would have been seen as a waste of divine intervention, and surely many ancient and medieval rulers also believed in the heavenly signs. If the astronomical symbols were strictly ornamental or strayed too far from the appearance of the actual event, then the design would have been of little benefit. Analysis has shown that acceptance of coinage was required for economic success, and economic success meant political success.

This book is about a beginning in understanding the use of astronomical symbols on coins in relation to political propaganda in ancient and medieval times, and the decision making processes of sovereigns that goes beyond the purely political considerations common to modern analyses. The combination of astronomy and numismatics is a starting point for additional research. With modern computers we can reconstruct the heavens; we can place ourselves back in time and imagine the awe of ordinary people who had no knowledge of the science of the heavens, but were swayed by pagan and religious superstitions. Research needs to be done to analyze the historical record in light of this new concept. Through coinage, humanity has preserved another source of information in a medium that has withstood the effects of nature over many centuries. Where paper may have disintegrated, metal has survived.

Capitalizing on celestial events as political propaganda on coinage was not confined to western cultures. Evidence of astronomical symbols on the coinage of Byzantine and eastern cultures also exists. A complete analysis of astronomy and numismatics would require a review of the use of such design symbols from ancient coinage to medieval, accompanied by a comparison of cultural migration with design utilization. Correlation of western astronomical design with that of eastern cultures could help to pinpoint common observations of astronomical events, especially those involving comets, planetary conjunctions, and solar observations along eclipse paths.

As the historical record is improved, one coin at a time, the ability to further understand the symbols will improve. It may be possible to solve many of the mysteries of coinage attribution, and thereby refine the historical understanding of entire regions as pieces of the puzzle fall into place. Much work remains to be done.

From the desire to influence the common man, we have inherited the knowledge of ancient and medieval thought, from the hammering of a die into a small piece of metal, one blow at a time.

Appendix A: Accuracy of Analyses

Medieval chronicler

The concept of astronomical interpretation and dating of ancient and medieval coinage based on the symbols used depends on accurate historical and astronomical data to support the analyses. Both types of data are subject to error, and the relative accuracy and significance of errors must be analyzed to assess the strength of the conclusions drawn from the data.

Historical Accuracy

Although some written documents exist and are referenced as supporting documentation for the astronomical events, a variety of error sources exist. The most obvious historical data error is one of incorrect recording of these events. For example, William of Malmesbury accurately described the total solar eclipse of 1133, but the *Anglo-Saxon Chronicle* stated the year as 1135 to coincide with the year Henry I died, possibly for dramatic effect. Florence of Worcester described the same eclipse as occurring in 1132.

A comparison with the records of other events with known dates permits the correction of such errors. Without such comparisons, historical errors in the year of an event may go unnoticed. Furthermore, the month and day of an event is more likely to be correct than the year. Medieval historians often predated their journals years in advance and may have recorded an event in the wrong year, have failed to correct the year annotations when lengthy descriptions took more than the allocated space for that year, or have failed to notice a blank year and then erroneously predated events for subsequent years. Later copyists might either have failed to correct such errors, or erroneously attempted to correct inconsistencies, and thereby perpetuated other errors.

Furthermore, some medieval historians began the new year at Christmas, others at the Annunciation on March 25, and still others in autumn. Even the time of day is difficult to interpret. Accurate clocks were not available, of course, but the time of day was often qualitative rather than numerical, such as using the term "mid day." William of Malmesbury described events by the hour of the day, which is roughly six hours earlier than what is now called mean solar or Universal Time. Presumably he was referring to the number of hours after sunrise.

Therefore the significance of historical data is to determine if an event were observed at all, and if so, to determine its historical significance. One must turn to modern calculations to determine precise chronology.

Numismatic Accuracy

This volume has presented many examples of ancient and medieval coinage with astronomical symbols. Some of the coins are fairly common and well documented over years of numismatic research. Other examples may either be extremely rare or considerable uncertainty may surround their attribution. Descriptions of coins are often based on references that were compiled more than a 100 years ago. For coins that are not common, descriptions may be based on a few coins that are badly worn. As the dies that were used to strike the coins were all hand made, often by different moneyers, minor variations may exist for the same coin type. In addition, astronomical symbols on coinage design were usually not of primary importance to numismatic researchers, and descriptions of secondary design features may not always be accurate.

For example, the primary reference for medieval Spanish coinage is *Monedas Hispano-Cristianas*. This highly respected, three-volume work was written by Aloiss Heiss, and the volumes were published in 1865, 1867, and 1869. Until the second half of the 20th century, when photographic illustrations became commonplace, numismatic illustrations were found as line drawings. Usually these drawings, along with descriptive narrative, are sufficient to identify a particular coin. However, consider the figure by Heiss for a dinero of Alfonso VIII of Castile. The reverse of this coin is illustrated in his reference on plate 4, number 18 (figure A1). For the narrative description of the reverse on this coin, Heiss writes, "encima del castillo, a la drecha, media luna; a la izquierda una estrella."

Figure A1. Heiss, plate 4, no. 18

Heiss has illustrated the reverse of this coin as a castle with a crescent moon to the left and a star to the right, although the translation of the narrative reverses the positions of the star and crescent. As figures A2–A3 clearly show, the star is actually a mullet. Does this mean that other examples exist with a star instead of a mullet? Or was the coin Heiss described an example that was worn so that the mullet appeared as a star? Most likely, it was not important to Heiss to distinguish between a star and a mullet. This is substantiated by his description of another variation (figure A4) of this coin: "encima del castillo, a la izquierda una [es]trella; a la drecha una B."

Figure A2. Actual coin   Figure A3. Mullet detail

The narrative position of the star and the *B* correctly match the illustrated figure, but note that in this case the star is shown as a mullet.

Figure A4. Heiss, plate 4, no. 16

For the purpose of astronomical analyses, a star may be a star, the sun, or possibly even a comet. A mullet, however, is more likely to represent a total solar eclipse. Thus, relying on standard references for precise descriptions of coins is sometimes difficult.

Astronomical Accuracy

The accuracy of determining astronomical events has two components. The first deals with the dynamical modeling of the mechanics (motion) of celestial bodies. With the sun as the origin of this reference system, then the motion of the planets, and in particular, the planets visible to the naked eye, and the relative motion of the Earth and the moon, can be fairly well modeled by analytical methods. The limiting factors in these models is the level of detail presented in considering minor perturbations to planetary and lunar motion and the extent to which computer analyses numerically process the models.

Numerical computations of the motions of the planets are fairly accurate over an astronomically short period of a few thousand years, and given the large distances between the planets, small errors in position do not significantly alter relative planetary positions as viewed from the Earth. Thus the astronomical interpretation of coinage design based on planetary conjunctions would have insignificant errors. Also the location of an observer on Earth is not particularly significant.

The relative positions of the sun, the Earth, and the moon are much more important when considering solar and lunar eclipses. Given the relatively small astronomical distance between the Earth and the moon, errors in calculating lunar motion could make the difference between an eclipse being total (annular) or partial. Furthermore, the path of totality (annularity) of a solar eclipse could be altered.

To determine local conditions for each eclipse, I used multiple sources to generate information for the material presented in this book. As a starting point, I employed Oppolzer's Canon of Eclipses to identify candidate events. Although written in 1887 (reprinted in 1962), the eclipse data is accurate enough to permit a good approximation of eclipse paths. Oppolzer's charts of totality and annularity are based on curves drawn through three or four computed points. The curves, however, do not account for the Earth's rotation and may therefore be in error by a few hundred miles.

Modern computer software provides an efficient way to predict eclipse paths accurately. However, the solar and lunar positions, as well as the rate of rotation of the Earth, are only approximately known for events that took place 1,000 years ago. This presents somewhat of a dilemma in determining an accurate path of totality or annularity. While a variation of a few percent in the lunar obscuration of the solar disk when considering a partial eclipse is insignificant, whether or not a particular location is in the path of totality or annularity may be extremely important.

To provide some degree of assurance of my results, I used four commercially available computer programs to analyze the eclipse events. Each program uses a somewhat different technique, and all the software developers claim a high degree of accuracy. The software programs agree within 1 percent on the magnitude and location of eclipses in a given location.

Given that most coin designs are regional, that is, a mint or province may cover a region that extends over 100 miles or more, the problem of finding the exact path of totality or annularity is minimized. However, two sources of error require further attention. The first concern considers whether or not the solar and lunar alignment with the Earth resulted in totality. While contemporary writings referred to many of the events cited, they may not have described whether an eclipse was total or partial.

An example of possible modeling or computational error is presented by the lunar eclipse of January 3, 1200. In *Images of History*, by Ralph of Diceto, dean of St. Paul's Cathedral, London (c.1180–1201), he describes the event as follows: "An eclipse of the moon occurred on 3 January in the middle of the night, for a duration of three hours. Presently it turned the colour of blood and emitted rays like fire."

While the "emitted rays like fire" is probably allegorical, the moon does turn red during a total lunar eclipse, when only the red wavelengths of sunlight are refracted by the atmosphere of the Earth and illuminate the lunar surface. All the text and computer software eclipse sources, however, predict only a partial lunar eclipse as depicted in figure A5.

While care must be taken in adjusting the lunar motion theory based on one historical piece of evidence, one can analyze what effect this adjustment might have. When the moon is in the penumbral area of the Earth's shadow, the moon's appearance changes little, but as the moon passes into the umbral region, the lunar disk appears to be obstructed. Depending on the geometry of the eclipse, the umbral phase can last for several hours.

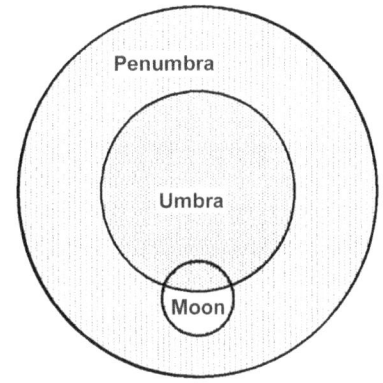

Figure A5. Lunar eclipse of January 3, 1200

The recorded three hours for the eclipse does not help to determine the geometry of the eclipse. If the eclipse had indeed been total, then the non-total umbral phases could have lasted for more than three hours. However, the longest period for full lunar totality within the umbral region would be less than two hours. Thus the three-hour duration does not help to determine the eclipse's geometry.

For totality to have occurred, the moon would have had to have been between 0.3° and 1.2° higher in the sky relative to the solar–Earth alignment (ecliptic). This would be a significant adjustment to the current lunar theory. However, if the moon were too low relative to the ecliptic, then the moon would be too high when it was between the Earth and the sun. Thus solar eclipse paths would have to be adjusted to the south. Such errors in lunar orbital modeling could result in eclipse path errors up to 100 miles or so.

Lunar motion theory is fairly accurate for about the last 20 centuries, but computational approximations and methods may have resulted in small errors that have combined in some manner to produce this large an error. If this were the case, then computational error probably varies from one eclipse to another, and this large an error is probably coincidental. In actuality, the lunar eclipse may have been partial, but was recorded as a total eclipse with "emitted rays like fire" for dramatic effect. At any rate, this evidence indicates that some north-south error in the eclipse paths is possible.

Another, and more likely, possibility is a date error in the written historical record. Modern computations predict that the maximum partial coverage of the lunar disk occurred at 5:25 in the morning on January 3, not in the middle of the night. But on June 28, 1200, a total lunar eclipse did occur shortly after midnight, and the umbral phase lasted for about three hours. Perhaps the written record of these two eclipses became combined into a single event.

The second area of concern is independent of the correctness of orbital motion, but deals with the irregularity and uncertainty of the rotation of the Earth, that is, what location on the Earth allows witnessing an event. While a lunar eclipse is visible from half of the Earth, Earth rotation error would only be significant if a location were at the edge of the area of visibility. The path of totality or annularity of a solar eclipse is extremely narrow, however, and uncertainty in the rotation of the Earth may become significant. Time relative to planetary (and lunar) motion is referred to as Dynamical or Ephemeris Time, while time on the Earth is referred to as Universal Time. Ephemeris Time is not affected by changes in the Earth's rotation, but Universal

Time is. The difference between these two time measures is referred to as Delta-T. The value of Delta-T must be determined to correctly describe the path of totality.

Astronomers have established empirical values for Delta-T based on recorded observations of eclipses and other astronomical events. Of course, the further back in time the event occurred, the less reliable the historical records, and therefore the values for Delta-T are also less reliable. Most sources use values based on research conducted by Stephenson and Morrison (1984). In 1997 Stephenson published new values for Delta-T based on additional research. Data points from a few recorded ancient eclipse events were chosen in addition to modern records to construct the Delta-T model.

Meeus (1991) fit a quadratic curve (parabola) to data from this research to provide algorithmic estimates of Delta-T for medieval and ancient times. Many computer analyses of astronomical events use the Meeus approximation when values for Delta-T are required. The physical mechanisms that cause the Earth's rotation rate to change will result in a quadratic function overlaid with short-term and long-term periodic variations.

As figure A6 shows, Delta-T values provided in the *Astronomical Almanac* (1991) from 1620 to the present, based on Stephenson and Morrison's research, do not fit a quadratic function, but appear to be the result of the interaction of several periodic effects that when combined cause high and low points in the values.

The research that formed the basis for the quadratic model also used a few ancient records of solar eclipses to help define the model. Ancient solar eclipses include those recorded in (or by) Babylon (1063 B.C.), Nineveh (763 B.C.), Archilochus (648 B.C.), Thales (585 B.C.), Pindar (463 B.C.), Thucydides (431 B.C.), Agathocles (310 B.C.), Hipparchus (129 B.C.), Phlegon (A.D. 29), Plutarch (A.D. 71), and Theon (A.D. 364). The dates associated with each eclipse are currently thought to be the most likely candidates for each event. Of these eclipses, the two best known are those of Thales and Nineveh (Amos 8:9).

Consider the eclipse recorded by Thales of Miletus (624–548 B.C.). Thales, an Ionian Greek mathematician and astronomer, was credited 100 years later by the Greek writer Herodotus to have predicted the eclipse that stopped the war between the Lydians and the Medes. During a battle in northern Turkey, the two sides immediately stopped fighting when the eclipse occurred and took the eclipse as a sign that they should make peace.

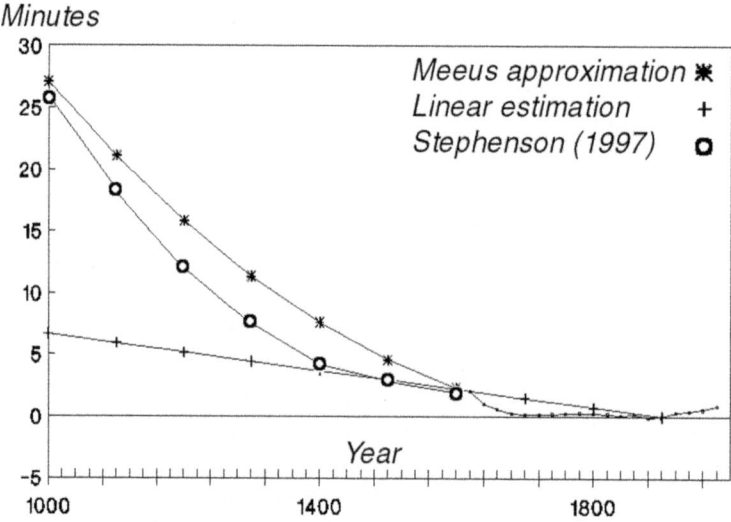

Figure A6. Values of Delta-T, 1000–1991

Modern science historians argue whether Thales could have predicted the eclipse. Nevertheless, northern Turkey is used as a position in the path of totality for the total solar eclipse of May 28, 585 B.C., and thus as a data point in the Delta-T quadratic model. However, is the 585 B.C. eclipse the correct one that ended the war? Given inaccuracies of a few years in the written records of medieval times, what is the potential for dating errors in ancient times? Thales lived between 624 and 548 B.C. and the war between the Lydians and the Medes lasted for several years. Other total solar eclipses in 610, 604, and 575 B.C. may have crossed northern Turkey during his lifetime and are possible candidate eclipses for the one that stopped the battle.

Only a few data points were used to compute the values for Delta-T in medieval and ancient times, and these data points are a continual source of research and discussion. Many of the medieval coinage examples discussed in this book seem to better fit a model with a slightly lower value for Delta-T. Thus for the eclipse analyses used herein, a linear estimation of Delta-T was used. This estimation is based on a straight line approximation between the zero point of Delta-T around the year 1900 and the value given in the *Astronomical Almanac* for 1620. The basis for using the linear approximation is that strong evidence indicates that friction from tidal actions slows the rotation of the Earth by a small amount each year, and this effect would be cumulative over the years. Other periodic changes to the Earth's rotation rate would then be adjustments to the cumulative tidal effect. To reduce any error in extrapolating the value of Delta-T back to medieval times because of short-term periodic effects, a very long baseline (1620–1900) was chosen.

While this results in about a 20 minute Delta-T difference relative to the Meeus approximation for the year 1000, the geographic difference amounts to an east-west shift of an eclipse path by about 200 miles at European latitudes and only about 60 miles by the year 1300.

Given that some minor north-south path error may be caused by lunar modeling or computation, differences resulting from the Delta-T model are not greatly significant, especially when considering a regional analysis of the event. The lower values of Delta-T do, however, seem to better fit the symbols found on medieval coinage. Therefore given all the positional uncertainties, not to mention non-viewability because of weather, astronomical event correlation must be based on historically recorded events as well as on predicted astronomical events.

All the eclipse paths shown in this manuscript were computed prior to 1995 as part of an earlier self-published work by the author, *Symbolic Messengers of Medieval Man*. The computer software programs used to generate the eclipse paths allowed input values for Delta-T computed from the linear estimation, and these results were compared with paths using the standard Meuss approximation algorithm from 1991. Therefore the Meuss eclipse paths shown in various figures in this manuscript reflect the values developed by Stephenson and Morrison in 1984. The new values for Delta-T published by Stephenson in 1997 are also plotted on figure A6 for comparison purposes. Note that the new values shift the Delta-T curve closer to the linear estimation values.

Accuracy of Ancient Predictions

Although the differences in eclipse paths predicted by the linear and quadratic models for Delta-T for the medieval era are not significant, analyses of ancient physical events are greatly affected. Positional accuracy for the sun, moon, and planets also affects eclipse and conjunction computations, but the theories that predict their positions are fairly well known. However, the path of totality or annularity of a solar eclipse is not very accurate, and analyses must consider such paths to be regional rather than precise. Analyses of astronomical symbols on ancient coins appear to favor paths predicted by the quadratic model even though the correct model for Delta-T is certainly very complex.

APPENDIX B: ADDITIONAL EXAMPLES

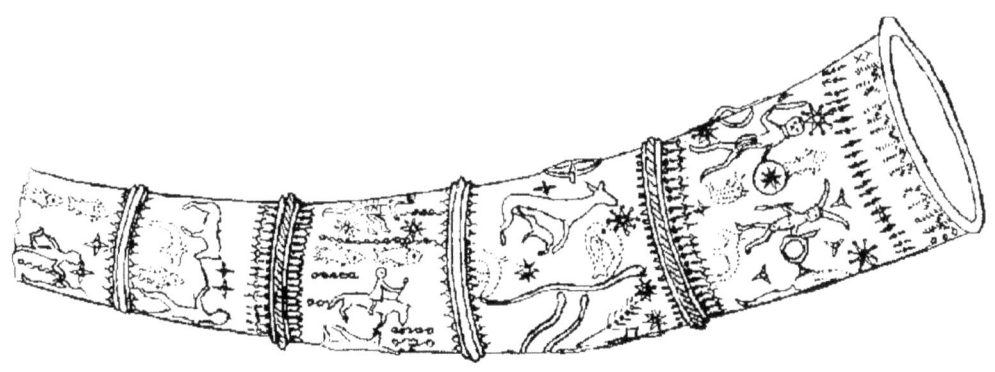

  The Gallehus gold horns are the most fantastic and attractive Danish finds of all from ancient times. The two horns are decorated with small vignettes containing numerous mullets and probably have religious significance. Several theories have been proposed to explain the meaning of the vignettes. One of the explanations is that they show a parade celebrating the solar eclipse that crossed Denmark in 413 A.D. The two horns were found near Gallehus, Denmark, in 1639 and 1734, and were stolen and melted down in 1802. Archaeologists date them to about 450.

190 Appendix B

Examples of astronomical symbols on ancient and medieval coinage number in the hundreds, perhaps even thousands. Obviously no single treatise can describe them all, much less analyze their significance. The previous chapters provided examples that demonstrated the basic concepts of depicting such symbols to represent both generic and actual celestial events. Chapters 6–9 showed how the numismatic record could add to historical knowledge. This appendix presents additional examples that either did not fit well into the context of the previous chapters or were more than was required to demonstrate the discussion points. Nevertheless, they are of interest in their own right, and are presented here for those readers who would like to see additional examples of astronomical symbols on ancient and medieval coinage.

Roman Republic

The star above the prow of a ship as an omen of victory was a popular motif. In 40 B.C. a denarius of Mark Antony was issued with the name of Ahenobarbus in the reverse legend and a star above the prow of a ship (figure B1). Ahenobarbus had won a decisive naval victory on the day of the first Battle of Philippi in 42 B.C. Ahenobarbus had followed Brutus to Macedonia after the assassination of Julius Caesar in 44 B.C. and was given command of a naval fleet. The armies of Mark Antony crossed the Adriatic Sea during the summer of 42 B.C., and the first Battle of Philippi was fought on October 27, 42 B.C.

Figure B1. Mark Antony and Ahenobarbus

After the suicide of Brutus at Philippi, Ahenobarbus became a pirate, but reconciled with Mark Antony in 40 B.C. A month before the first Battle of Philippi, on September 21, 42 B.C., a total solar eclipse crossed central Africa and would have been seen across northern Africa as a partial eclipse. Roman legions occupied much of northern Africa and knowledge of the eclipse would have made its way back to Rome. Was this eclipse taken as an omen for the victory at Philippi and does the star above the prow of the ship refer to his event?

The Annular Eclipse of 1147

The 1147 eclipse was of moderate size and duration. The path of annularity ranged from 196 to 253 kilometers in width (figure B2). Annularity occurred between 9:30 and 12:06 UT. At its maximum view, annularity was visible for 4 minutes and 11 seconds. The viewing conditions must have been quite favorable, as examples of coinage designs that refer to this eclipse exist all along its path. Observation of the eclipse was recorded in England, Holland, France, Germany, Switzerland, and Byzantium.

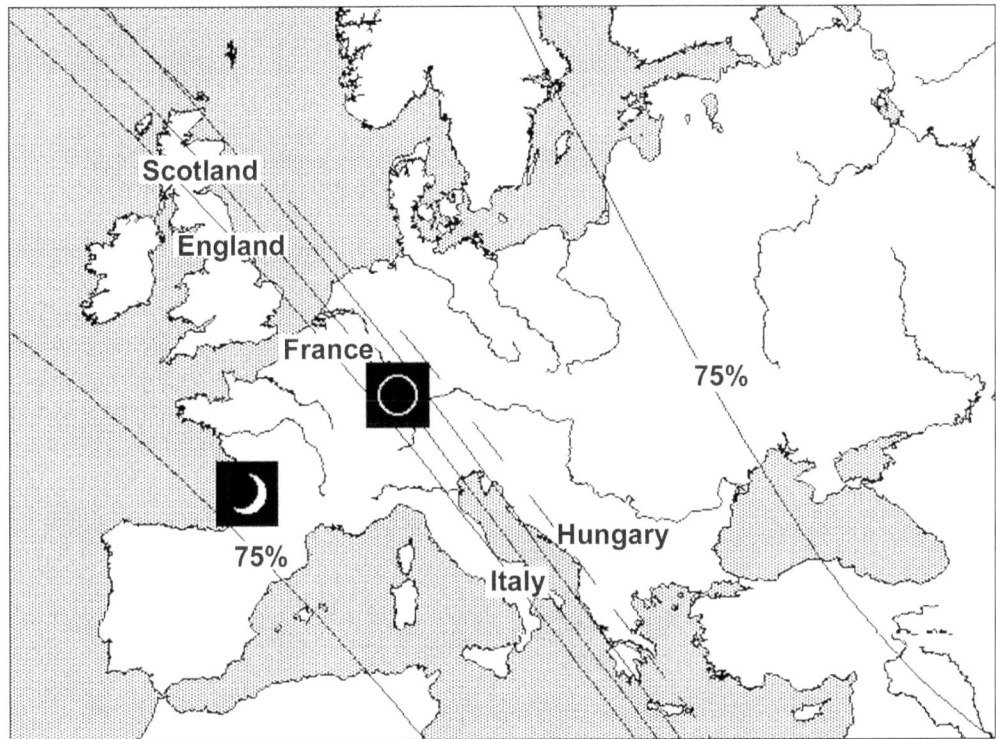

Figure B2. Path of the October 26, 1147, annular solar eclipse

Scotland

David I of Scotland struck an extremely rare variety of his penny in the mid to late 1140s that contained either crescents or annulets with a central pellet in the quadrants of the reverse cross (figure 418). Another extremely rare penny replaces single pellets with a sun symbol composed of annulets in the second quadrant and a star in the third quadrant (figure 419). A similar penny with single pellets only has a star at the end of the reverse legend.

England

Much of the regular and irregular coinage struck during the 1138–1153 anarchy in England contains stars, mullets, and annulets. A variety struck in the Midlands, possibly during the Interdict of 1148, has a central star and four annulets (figure 415).

Feudal French Provinces

Across the English Channel in the feudal French province of Ponthieu, Jean I (1147–91) issued a denier with annulets commemorating the eclipse that occurred in the first year of his reign (figure B3).

As early as 1148 in Champagne, Archbishop of Reims Samson de Mauvoisin issued deniers with crescents in the quadrants of the cross, presumably representing the appearance of the sun near the 1147 path of annularity. This design type became immobilized in Reims until the middle of the 14th century (figure B4).

Figure B3. Jean I

Figure B4. Denier of Samson de Mauvoisin

In Artois, Anselme, the count of Saint-Pol (1150–74), issued a denier with annulets in two of the reverse cross quadrants and an inverted *V* or pyramid at the base of an oat stalk (figure B5). The pyramid may represent Halley's comet of 1145 or the comet of 1151, and the annulets surely represent the annular eclipse of 1147 that crossed Artois. Comets and eclipses were often associated with the weather, including storms and drought. Perhaps the symbols had agricultural significance.

Figure B5. Denier of Anselme

Petite deniers of St. Omer similar to those with mullets representing the 1135 total solar eclipse are also found with annulets. In the Bethune region of Artois, petite deniers are found with annulets alone or with mullets, and the coins most likely refer to the 1147 eclipse (figures B6–B7).

Figure B6. St. Omer (1128–68)?

Figure B7. Bethune: Robert IV (1145–92)?

Similar astronomical designs are found on coins of Picardy, just to the south of Artois. Early issues of Foulques II of Amiens (1031–58) have a crescent in the first quadrant of the reverse cross (figure B8), perhaps in reference to the appearance of the 1039 solar eclipse that was seen off the coast of Brittany and would have appeared as a thin crescent in Picardy, or in reference to the 1147 eclipse.

Figure B8. Denier of Foulques

Sicily

Roger II was the count of Sicily from 1105 until 1130, when he had himself crowned king at Palermo and reigned until 1154. Early coins of Roger II appear to be of Arabic, Byzantine, or various original designs. As king he issued small coins (tari, quarter-follis) without annulets. Any early issues of Sicily containing annulets would have been immobilized from the annular eclipse that crossed Sicily in 1098. In 1140 Roger II ordered a coinage reform. He forbade the use of the early Norman romesini copper coins, and replaced them with the ducalis (ducatus), derived from the duchy of Apulia, which Roger II had annexed in 1127.

The ducalis was Byzantine in design, and was intended for use on the mainland, where strong Byzantine traditions prevailed. The third-ducalis, struck at Palermo, had both Latin and Arabic inscriptions, was intended for use in Sicily, and contains an annulet in the design (figure B9). Subsequent kings, such as William II, immobilized the annulet in their designs. The annular eclipse of 1147 crossed directly over Apulia (figure B10), therefore third-ducalis coins with annulets would have been struck after October 1147.

Figure B9. Third-ducalis

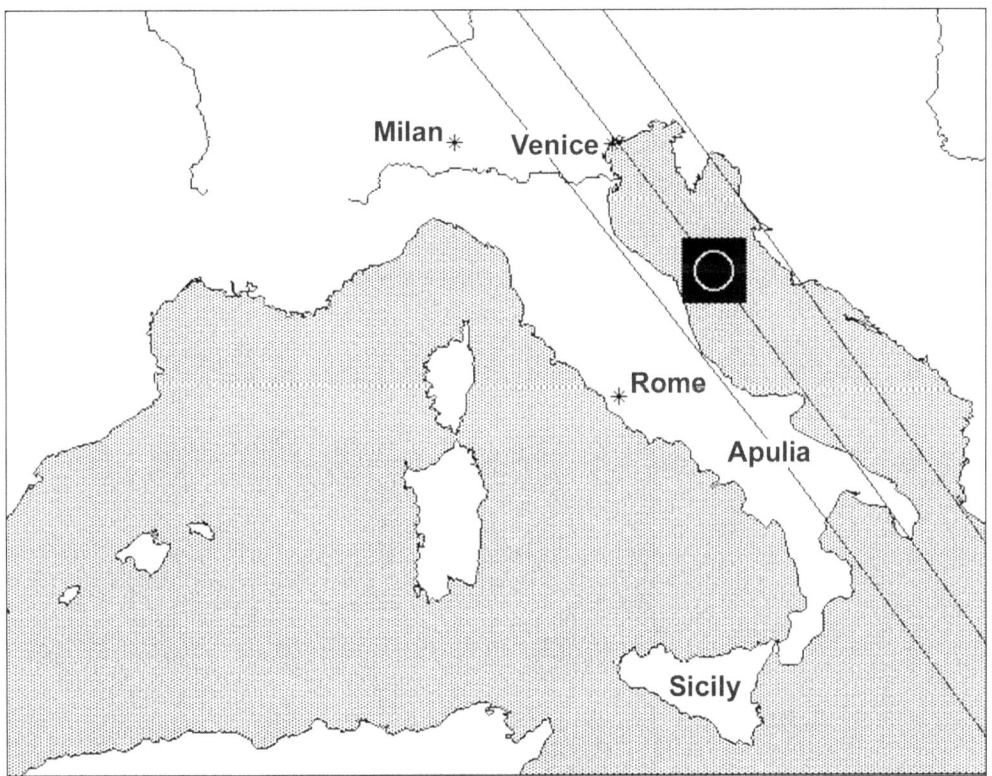

Figure B10. Path of 1147 annular eclipse over Apulia

Hungary

During the second half of the ninth century Magyars invaded the Carpathian Basin from the Khazar state north of the Black Sea. Little is known about them prior to this southwestern migration to the middle Danube River region. The Magyars were made up of seven loosely organized, semi-nomadic tribes led by the elected chieftain, Arpad. During the 10th century, defeats by forces of the Holy Roman Empire changed these tribes, which moved them away from paganism and adopted a more western lifestyle. In 972, Prince Geza became leader of the Hungarian confederation and allowed his subjects to convert to Catholicism. His son, Stephen, worked hard to eradicate paganism, and in 1000 the pope rewarded him by naming him king of Hungary.

Over the next three centuries Hungary acquired new territories and a multinational, multilingual character as a result of these acquisitions and of immigration by foreign colonists who occupied uninhabited areas. Although many of the medieval Hungarian issues are quite crude, the designs on these coins are rich with depictions of contemporary beliefs and contain many astronomical symbols. Beginning with the reign of Stephen, the need for an unquestioning obedience by the population and the supremacy of royal authority was required to consolidate the loose confederation of tribes. The use of astronomical events as a propaganda device to uphold the monarchy as a divine right is certainly probable. Given the kingdom's multi-lingual nature, widely accepted symbols would have played a more significant role than written characters. Many of the coins of the Arpadian kings contain no legends at all, only pictorial devices and symbols.

In the first half of the 12th century, the paths of total solar eclipses grazed the eastern and western Hungarian borders in 1115 and 1133, respectively, and the 1147 annular eclipse traversed nearby over the Adriatic Sea. The relationship to coinage designs is too unclear to permit any definite conclusions; however, crescents enclosing pellets often played a predominant role in coinage design during this period, and may be the basis for later design immobilization. On some of the issues, three or four small triangles are associated with one of the crescents. The example shown in figure B11 has four triangles, suggesting an issue in the middle of the twelfth century, as four comets were recorded between 1106 and 1145, although the triangles on this coin may be the mark of a moneyer.

Figure B11. Hungarian denar

The Annular Eclipse of 1207

The annular eclipse of 1207 followed a similar track to that of the one in 1153, but was more moderate in size and duration. The path of annularity ranged from 221 to 292 kilometers in width (figure B12). Annularity occurred between 9:38 and 12:22 UT. At its maximum view, annularity was visible for 4 minutes and 32 seconds. As the practice of using actual astronomical events on coins spread across Europe, examples abound of designs that may have represented

Additional Examples 195

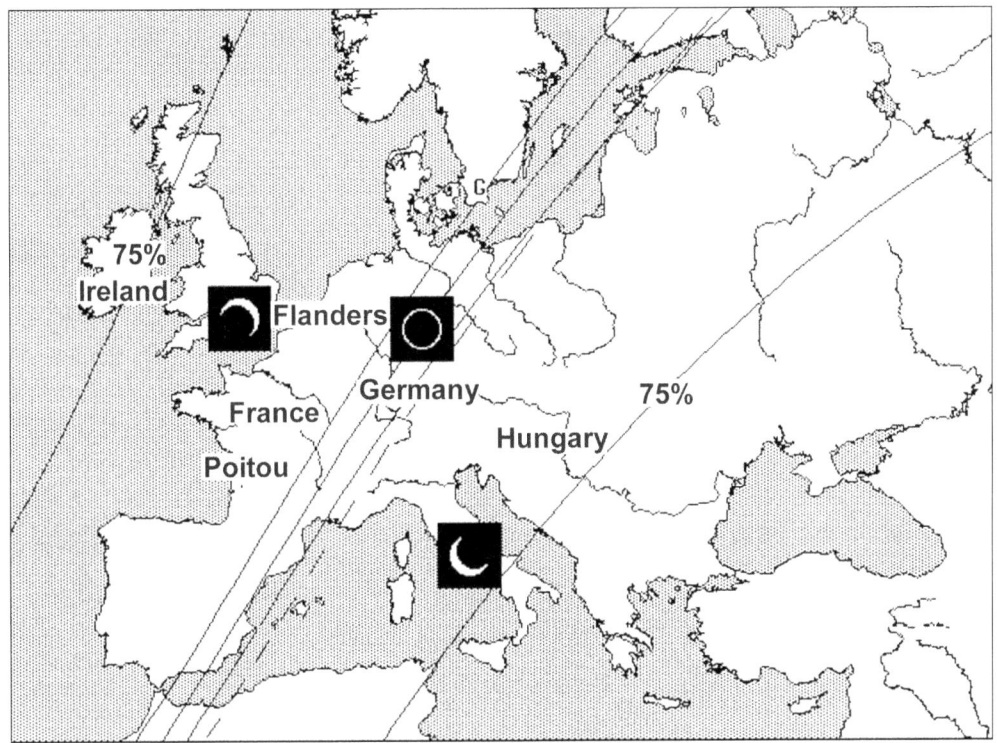

Figure B12. Path of the February 20, 1207, annular solar eclipse

this eclipse. Contemporary chroniclers in England, France, Germany, Holland, and Estonia recorded the eclipse.

Ireland

In English-controlled Ireland, King John issued his third set of coinage around 1207 which depicted a crescent and the sun (figure B13). Both the eclipses of 1201 and 1207 are potential sources for the design.

Figure B13. Ireland

Poitou

Although not minted in the name of King John of England, his vassal in the French feudal region of Poitou, Savary de Mauleon (1199–1216) issued "METALO" (a spelling for the Melle region of western France) deniers with a crescent (figure B14).

Figure B14. Poitou

Feudal French Provinces

In the nearby French province of Berry, William I of Deols (1203–33) replaced the central pellet with a crescent (figure B15).

Figure B15. Deols

Flanders

Farther to the north, in Flanders, a petit denier was struck (c. 1220) that contains alternating crescents and annulets around the obverse field containing a fleur-de-lis (figure B16).

Figure B16. Flanders

Germany

The cities of Mainz and Wurzburg were situated on opposite sides of the central path of the eclipse and both localities were within the path of annularity. Both mints issued coinage with annulets in the design. In Mainz, a bracteate of Detrich the Oppressed (1197–1221) has Margrave standing with a large annulet over a cross (figure B17). In the episcopal mint of Wurzburg, the coinage of Otto I von Lobdeburg (1207–23) has an annulet by a sword on the obverse and two annulets over the city on the reverse (figure B18).

Figure B17. Mainz   Figure B18. Wurzburg

Hungary

In the eastern European kingdom of Hungary, Andreas II (1205–35) copied the annulets and city gate design on his oboles and struck other coins with crescents and mullets (figure B19). The 1207 eclipse was seen in Hungary as a crescent with about 80 percent of the solar disk covered by the moon.

Figure B19. Andreas II

The Total Eclipse of 1239

The total solar eclipse of 1239 was of moderate duration. The path of totality ranged from 218 to 264 kilometers in width, with a maximum view of totality lasting for 5 minutes and 59 seconds (figure B20). Totality was viewable between 10:21 and 13:37 UT. This eclipse is a good example of the use of astronomical symbols on coinage across diverse cultures in Europe and Byzantium, and was recorded in Portugal, Spain, France, England, Germany, Italy, Austria, and Hungary. The 1239 eclipse was followed by another total eclipse in 1241 (discussed in chapter 3), and these two eclipses were the basis for numerous examples of solar symbols on medieval European coinage during the second half of the 13th century.

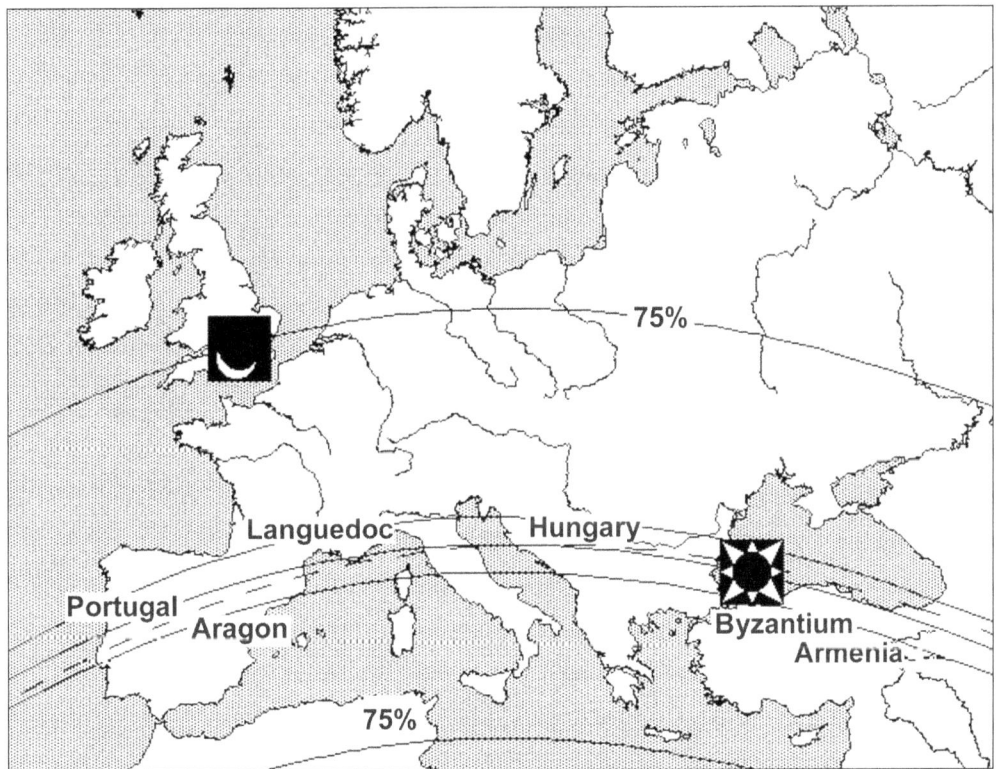

Figure B20. Path of the June 3, 1239, total solar eclipse

Portugal

The total solar eclipse that crossed the Iberian peninsula in 1239 may have been viewed in Portugal as a sign for the recapture of Islamic held lands. Although the dineros of Sancho II do not contain stars, one type has crescents in the quadrants (figure B21). Subsequently minted dineros of Alfonso III (1248–79) have a star or alternating stars and crescents in each of the four quadrants of a cross.

Figure B21. Sancho II (1223–48)

Aragon and Languedoc

James I of Aragon (1238–76) issued a denaro in Valencia that has crescents with three pellets (figure B22). Venus, Saturn, and Mercury may have been visible during the eclipse. Some numismatists think that the reverse is a floral design unrelated to the eclipse. However, in the Languedoc region of Montpellier, James I issued a gros with mullets in the reverse design (figure B23). The total solar eclipse of 1239 crossed directly over this area.

Figure B22. Valencia denaro of James I   Figure B23. Gros of James I of Aragon

Hungary

The beginning of the reign of Bela IV of Hungary (1235–70) was distinguished by two total solar eclipses, the 1239 eclipse and another in 1241. Mullets played both central and secondary themes in the designs on many of his coins (figures B24–B25). Some of the issues of Bela IV that have the mullet and crescent motif may be immobilized designs, but many are no doubt representative of the two total eclipses.

Figure B24. Bela IV   Figure B25. Croatia (Slavonia)

Byzantium

In Nicaea, John III (1222–54) modified a trachy by adding stars with a central pellet to each side of Christ standing on a coin (figure B26). Similar issues of John III in Thessalonica, a region captured by Nicaean forces in 1246, do not have the stars, suggesting that the trachy with stars may have been issued prior to 1246.

Cilician Armenia

A star with a central pellet (figure B27) was added to the base of the staff on trams of Hetoum and Zabel (1226–71). These trams are also found without the star, with pellets, or with annuletted pellets on the staff. In addition to the 1239 total solar eclipse, another total solar eclipse crossed this region in 1267, and an annular eclipse crossed it in 1263.

Figure B26. Nicaea

Figure B27. Cilicia

The Annular-Total Eclipse of 1321

This annular-total eclipse had presented the typical narrow path of annularity and totality, when the apparent diameters of the sun and the moon are almost equal. Annularity began over the western edge of the Iberian peninsula early in the morning at 5:15 UT, where the path of annularity started with a width of only 42 kilometers, and gradually narrowed to 3 kilometers by the time totality occurred at 5:35 UT, when the central path crossed Russia (figure B28). The path of totality increased to a maximum width of 24 kilometers, and then narrowed again until annularity once again prevailed around 8:00 UT, with the second annular phase ending some 17 minutes later.

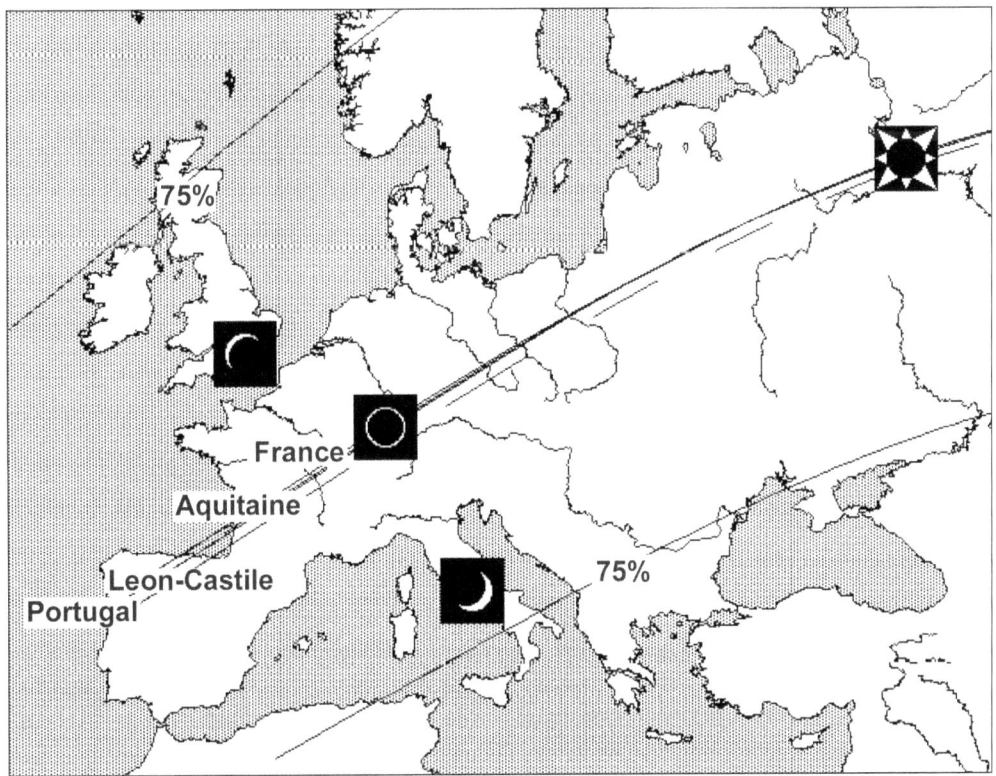

Figure B28. Path of the June 26, 1321, annular-total eclipse

At its maximum view, annularity was visible for only 29 seconds, and totality was seen at best for 27 seconds. Given the event's early morning time, coupled with its extremely narrow central path, it is not surprising that the use of symbols that might represent this event appears to be limited to those areas directly in or near the central path. The eclipse was recorded in Germany and Poland.

Portugal

In 1310, an annular eclipse crossed the northwest tip of Spain and was seen as a thin crescent in Portugal. The next annular eclipse to cross Portugal was the beginning phase of the annular-total eclipse of 1321. Some of the dineros of Denis (1279–1325) replace two of the stars in the quadrants of the cross with crescents (figure B29), and later dineros replace pellets on both sides of the *D* in the legend with annulets (figure B30). The annulets on coins of Denis are probably associated with the eclipses of 1310 or 1321. Total solar eclipses also crossed the ocean to the east of the peninsula in 1285 and 1295, and would have appeared as crescents in Portugal. Any of these events may be the basis for the crescent types of Denis.

Figure B29. Crescent dinero of Denis

Figure B30. Annulet type

The dineros of Alfonso IV (1325–57) have stars and crescents, or pellets and crescents, alternating in the quadrants of a cross (figure B31). The stars and crescents may be associated with eclipses of 1321, 1333, or 1354 that crossed Portugal.

Figure B31. Dinero of Alfonso IV

Leon-Castile

Alfonso XI, born in 1310, succeeded Ferdinand IV and ruled Leon-Castile from 1312 to 1350. From 1312 to 1324, his grandmother held the regency. Most of Alfonso's coins are without astronomical symbols; however, some astronomical varieties exist. His noven of Leon contains an annulet above a castle on the obverse, and to the side or at the foot of a lion on the reverse (figure B32). Some cornados have one or two mullets above or at the side of a castle (figure B33), and another has mullets on the collar of his bust.

During the regency period of Alfonso XI, his grandmother strived to save the throne for him,

Figure B32. Noven of Leon

Figure B33. Cornado of Alfonso XI

and propaganda to suggest his divine right to the throne would have been a useful tool for her. On June 26, 1321, the narrow path of the annular phase of the annular-total eclipse passed directly over Leon (figure B34), and was visible as a crescent elsewhere on the peninsula. There can be little doubt that the annulet issues of Leon symbolize this event. The annular eclipse of 1333 also crossed the region just north of Leon, and the variation of the placement of the annulet by the lion may reflect the first or second eclipse.

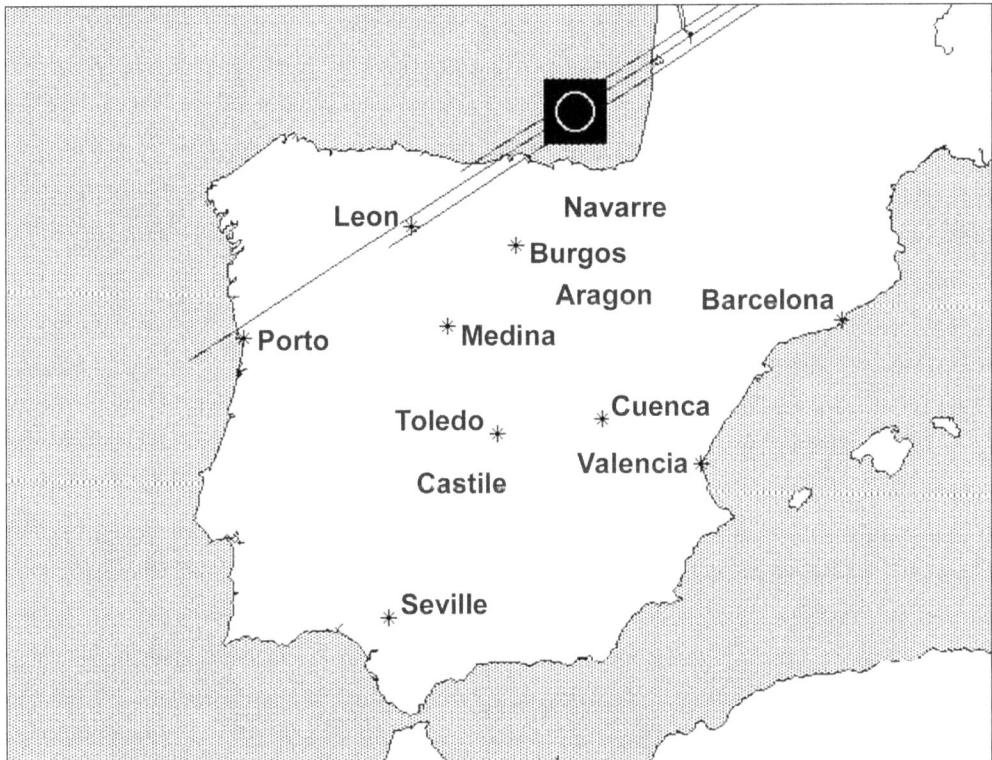

Figure B34. Path of the 1321 eclipse over the Iberian peninsula

Aquitaine and Royal France

The 1321 eclipse crossed Aquitaine and the Burgundy region of France (figure B35). Astronomical designs on coinage of Aquitaine were discussed in chapter 8. Beginning with Louis VIII in 1223, subsequent royal coinage of France contains astronomical symbols mainly as design stops in legends or as mint or moneyer marks. The limited use of these symbols on French coinage for

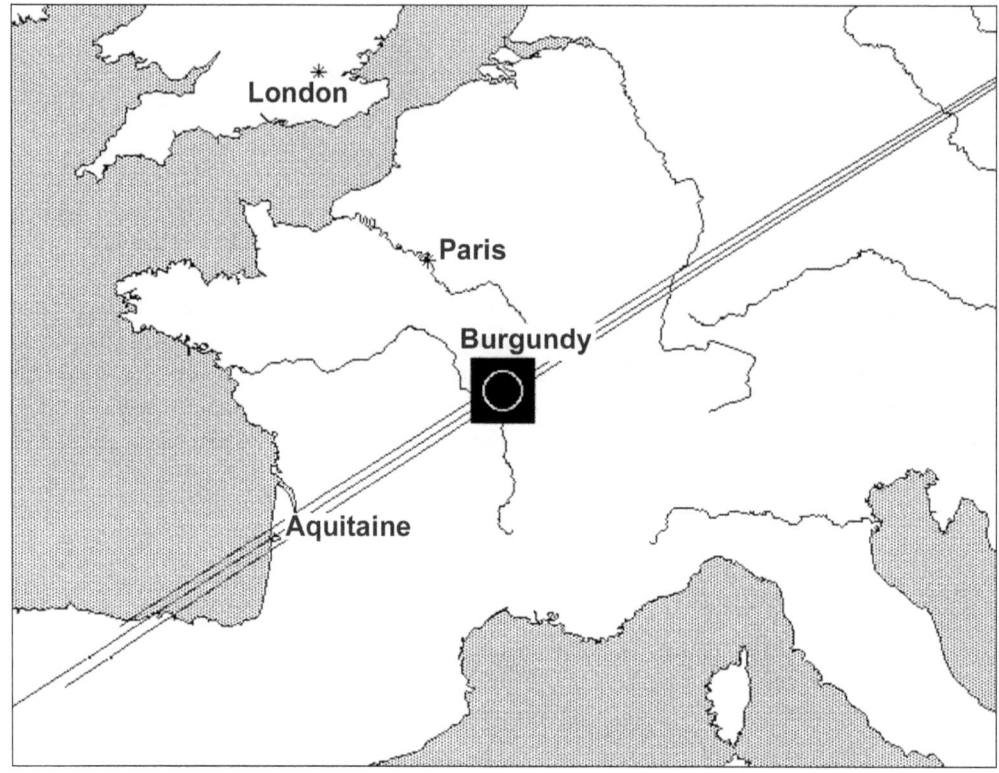

Figure B35. Path of the 1321 eclipse over Aquitaine and France

the next 100 years may have resulted from religious inquisitions that started in 1230. Several exceptions exist.

One example has annulets in the simple and double parisis of Charles IV (with immobilized similar types of Philip VI in 1328 and 1329). In October 1322, a parisis simple (figure B36) was issued with annulets on both sides of the *K* [Karolvs]. While the double parisis of 1322 does not contain an annulet, one was added during the third revision of the coin in 1326 (figure B37). The total-annular eclipse of 1321 crossed over the middle of France in its annular phase.

Figure B36. Parisis Simple   Figure B37. Double Parisis

In December 1339 a star was added to the obole tournois of Philip VI (figure B38). In July 1339 an annular-total solar eclipse crossed the northernmost tip of Scotland and the North Sea in totality, and cut through Europe from the Low Countries to Hungary as an annular eclipse. Possibly this eclipse was the basis for the addition of the star.

In 1355 Jean II struck an obole tournois with annulets on both sides of a temple transformed

Figure B38. Obole tournois

Figure B39. Obole of 1355

into a crown (figure B39), and on the gros a l'etoile struck in 1359, mullets were added in opposite corners of a cross (figure B40). The astronomical symbols are clearly significant to the design on both of these coins, rather than being decorative additions such as legend stops. On September 17, 1354, an annular-total solar eclipse crossed the southwestern Franco-Spanish border, with the change from annularity to totality occurring at or near the border region.

Figure B40. Gros a l'etoile

Feudal French Provinces

In Burgundy, Eudes IV (1315–50), added annulets (figure B41) to the immobilized design of previous dukes (see figure 162).

Figure B41. Denier of Eudes IV

Miscellaneous European Examples

The previous discussion demonstrated the use of solar symbols along eclipse paths. Additional astronomical examples are better described in the context of local events and history.

Ireland

In 1172 Prince John of England was given the lordship of Ireland and paid his first visit there in 1185. His first Irish coinage was a halfpenny (c.1185) with a right-facing bust on the obverse and a cross with lis and pellet in each angle on the reverse. This coin may have been issued by

John de Curcy, lord of Ulster. John's second coinage (c.1190–99) has a round moon-like bust on the obverse and a cross with annulets in each angle on the reverse (figure B42). This bust is certainly atypical of medieval coins. Could it represent "the man in the moon?" The annulets might represent an annular solar eclipse.

Figure B42. Second Irish coinage of John

Annular eclipses were visible in Dublin, Ireland, on January 28, 1180, and June 23, 1191. On May 1, 1185, a total solar eclipse was observable from just off the northern coast of Ireland and appeared as a very thin crescent in Dublin. Partial eclipses may have been seen on September 13, 1178, December 6, 1192, and November 27, 1201. The 1180 eclipse was probably too early for the possible dates of John's second coinage. In addition, annularity occurred when the sun was less than 3° above the horizon, and only the pre-annular partial phase may have been observed. Given that John made his first visit to Ireland in 1185, coupled with the rarity of observing the thin solar crescent because of the proximity to the path of the total solar eclipse, the moon-like face on the coinage may indeed be associated with the 1185 eclipse. However, the 1191 eclipse may be represented by the annulets on the coinage, implying that the second coinage was issued after June 1191.

The observation of the 1191 eclipse was documented in English chronicles. In Winchester, Richard of Devizes dismissed the superstitions of those who saw the event as a celestial omen as follows:

> Those who do not understand the causes of things marvelled greatly that, although the sun was not darkened by any clouds, in the middle of the day, it shone with less than ordinary brightness. Those who study the workings of the world, however, say that certain defects of the sun and moon do not signify anything.

The St. Patrick coinage of John de Curcy (c.1185–1205) has annulets with pellets on the reverse design (figure B43). Anonymous St. Patrick issues (c.1195–1205) have a cross with crescents in each angle (figure B44).

Figure B43. Coinage of John de Curcy

Figure B44. St. Patrick issue

Scotland

Malcolm IV of Scotland (1153–65), issued a penny with rosettes in the first and third quadrants of the cross (figure B45). Perhaps the rosettes are a solar symbol immobilized from the coinage of his father, David I (see figure 419) and a representation of the 1147 eclipse that crossed Scotland.

The early coinage of William of Scotland (c.1165–74) has a bust on the obverse with a cross on the reverse with either lis or crosslets in the angles. The lis and crosslets were replaced with a pellet within a crescent (c.1174–95), and perhaps a partial eclipse was the stimulus for the motif. Some varieties have the pellet and crescent behind the king's bust (figure B46). Bateson (1997) suggests an issue date as early as 1170. These crescent and pellet pennies have been grouped into two types: type I (c.1170/74–80) and type II (c.1180–95). Each of the two types is characterized by a distinctive sceptre that corresponds to the sceptres found on English Tealby (1158–80) and short cross (1180–89) pennies of Henry II.

Figure B45. Penny of Malcolm IV   Figure B46. Crescent and pellets

Both of the Scottish crescent and pellet types were struck at various mints in Scotland. However, William was captured in the unsuccessful uprising against Henry II of England in 1173, and following the Treaty of Falaise in 1174 he lost five castles including Edinburgh, Roxburgh, and Berwick, which housed three of his five mints. Many of the known Scottish pennies of the first type were struck in Perth, suggesting that limited mint operations continued in Scotland after August 10, 1175, when the castles were transferred to English control. Edinburgh was restored to William as part of the dowry of Ermengarde, whom he married in 1186, thereby reestablishing that mint under his control. The other two mints were restored in 1189.

Four solar eclipses were visible in Scotland between the beginning of William's reign in 1165 and his marriage in 1186. Partial eclipses occurred on April 11, 1176, and September 13, 1178. The paths of annular and total eclipses traversed Scotland on January 28, 1180, and May 1, 1185, respectively. As seen from Edinburgh, only about half of the solar disk was eclipsed during the 1176 event, but the eclipse occurred just as the sun was rising, and therefore the thickness of the atmosphere near the horizon may have made the event viewable. Mars was on one side of the solar crescent and Venus was on the other side. The 1178 partial eclipse was seen with more than 70 percent of the solar disk covered, and Venus was extremely close to the crescent. The 1178 and 1185 eclipses were recorded in Scottish chronicles, but the 1176 event went unrecorded throughout all of the British Isles.

Some of the type I pennies were struck at mints lost to England in 1175 and restored after 1185. Thus the type I pennies must have been issued prior to August 1175. This date precedes all the eclipse events and implies that a contemporary eclipse could not have been a stimulus for the crescent and pellet design. Two other possible explanations for the design should be considered.

On September 12, 1170, Jupiter and Mars not only came close enough together to be seen as a single star, but Mars actually passed directly in front of Jupiter. This is an extremely rare astronomical event. Gervase of Canterbury recorded this mutual planetary occultation: "On the Ides of September, at midnight, two planets were seen in conjunction to such a degree that it appeared as though they had been one and the same star; but immediately they were separated from each other."

Perhaps the crescent and pellet represent this celestial event. However, a pelleted annulet would have been a more appropriate symbol, but the event was seen and the date of the event fits the currently proposed issue date of the coinage.

The actual operations of the mints offer the other possible explanation. Written records of Scottish coinage of the time are very rare, and the dating of William's crescent and pellet coinage is based on sparse hoard data, knowledge of the castles being forfeited as part of the Treaty of Falaise, and design comparison with the coinage of Henry II. The *Chronicles of Melrose* mentions a new coinage in 1195, which is assumed to be William's short cross and stars coinage (figure B47).

Figure B47. Short cross and stars type

Forty years prior to the crescent and pellet coinage, the first coins of Scotland were struck in 1136 when David I captured Carlisle and its mint at the beginning of Stephen's reign. Some of the early coinage may have been struck in Carlisle prior to the opening of the Edinburgh mint, but with the mint name of Edinburgh in the coin legend. Perhaps some of William's type I pennies were struck at Perth or another location with the names of forfeited mints in a subtle act of defiance of the Treaty of Falaise. This would offer the possibility that the type I pennies were issued after either the 1176 or 1178 partial eclipse. It could also mean that William's type II pennies were struck soon after Henry's new coinage in 1180.

Based on the *Chronicles of Melrose*, the short cross and stars coinage of William is currently thought to have been issued in 1195, and continued to be struck for another 16 years after William's death in 1214. If the stars represent the total solar eclipse of 1185 (figure B48), which may have been taken as a heavenly sign for the restoration of Edinburgh to Scottish control, then these coins may have been issued as early as 1186. Even if the short cross and stars type were issued as late as 1195, the stars could still be a depiction of the 1185 eclipse.

Subsequent silver coins of Scotland from this period through 1390 all contain stars or mullets in the each of the angles of the cross on the reverse of the coin. Solar eclipses crossed Scotland in 1230, 1279, and 1330, and may have served as the basis to immobilize the design of William's coinage (figure B49).

The stars were replaced by mullets in 1280 during the reign of Alexander III, after 30 years of using stars on his coins (figure B50). The new coins closely followed the English re-coinage of 1279 with a new style, long cross on the reverse, along with much better workmanship on the coins. In April 1279 an annular eclipse crossed Scotland. The star may have been changed to a mullet to update the symbols based on the annular eclipse, or may simply have been due to the improved workmanship of the dies.

David II succeeded his father, Robert Bruce, in 1329 at the age of five, and was crowned in 1331. His first coins, which contained mullets, were issued between 1330 and 1333. A total solar eclipse in 1330 (figure B51) may have reinforced his divine right to the throne, especially because Edward Baliol challenged his right to rule.

In addition to Scotland, examples of mullets representing the 1330 eclipse that were struck by Edward III in England (and also on his Irish coins) and Aquitaine, Walrum of Julich in

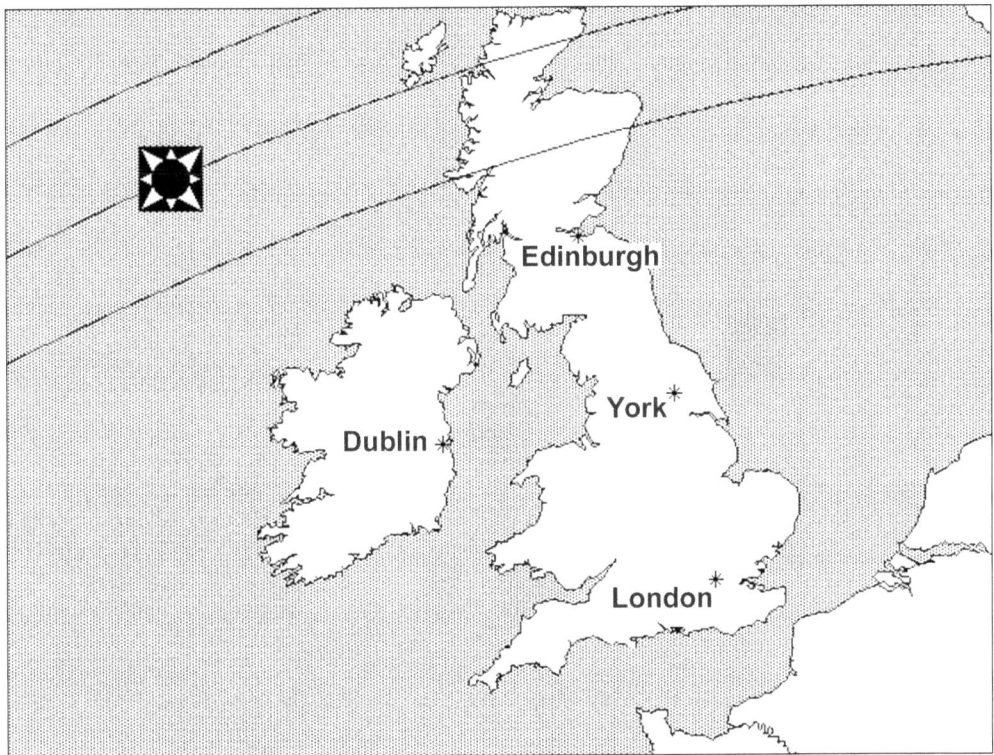

Figure B48. Path of the May 1, 1185, total solar eclipse

Figure B49. Alexander II, 1235–49   Figure B50. Alexander III

Cologne, and Raymond of Orange (feudal France) were discussed in previous chapters. Other coins of this period with stars or mullets were struck in Hungary, Bulgaria, and Byzantium.

England

As part of the 1279 re-coinage, Edward I ordered the first English groat (four pence) to be struck. Similar denominations had been introduced in France and Spanish territories during the preceding decade to facilitate Mediterranean trade and to pay for crusades. The Italian grosso had been in use since the beginning of the century. This first English groat was either experimental or not accepted, as no English groats were struck between 1282 and 1351.

An interesting feature of this groat is the mullet found both sides of the bust and on the breast of Edward I (figure B52). Others refer to the mullets as rosettes, but the design is similar to that

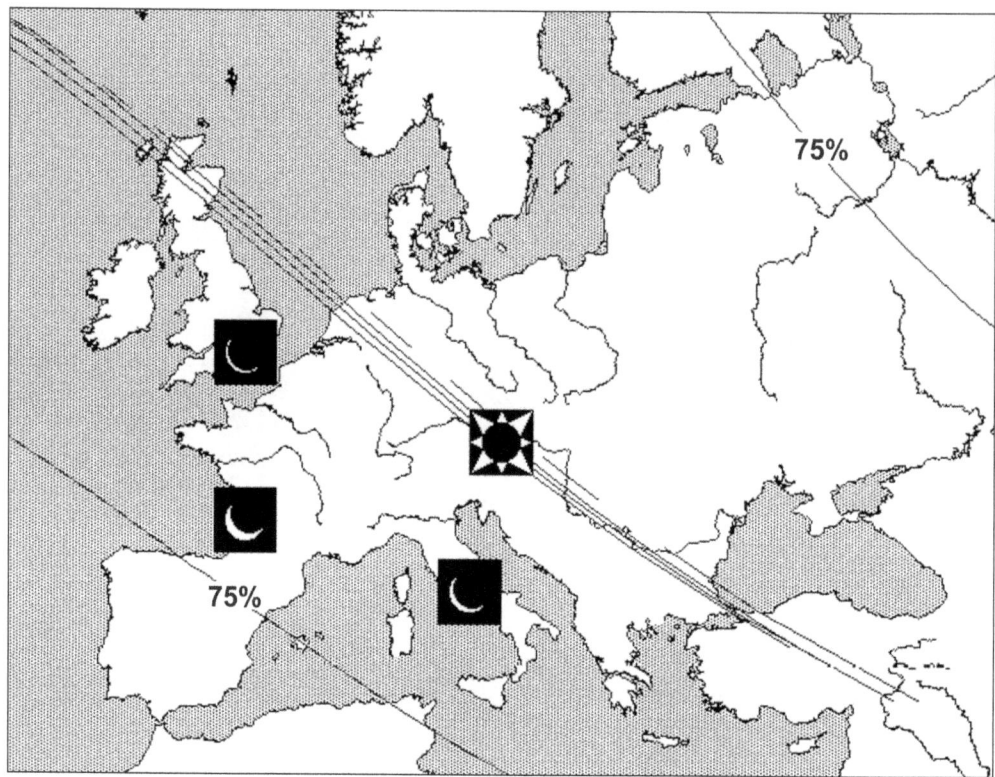

Figure B51. Path of the July 16, 1330, total solar eclipse

of Majorcan groats beginning with James II of Aragon in 1276 and continued by his successors, where the symbols are clearly mullets (see figure 181). The recognition of this symbol as being astronomical is further demonstrated by examination of English jetons that have mullets (figure 431) and most likely represent the 1279 eclipse.

Figure B52. Groat of Edward I

Feudal France

In Nevers, the use of the star in coinage design became immobilized starting with the issues of Herve IV of Donzy (1199–1223) through those of Louis de Dampierre of Flanders in 1321. The transition of coinage design leading up to the stellar motif is interesting (figures B53–B56). Issues from the middle of the 10th century until 1161 employed the name of Louis IV in the legend. The letters *REX* in the central field became degenerated over time, and by the time of Pierre de Courtenay (1184–99), the *X* had changed into a cross and the *RE* had been formed into a sickle. Herve of Donzy replaced the cross and one of the pellets with a star, perhaps in reference to a celestial event.

Figure B53. Obole in the name of Louis IV

Figure B54. Degeneration of REX

Figure B55. Pierre de Courtenay

Figure B56. Herve de Donzy

Venice

In 1202 a crusading army was gathering in Venice preparing to set sail. The city required a larger denomination of silver coinage to pay for shipbuilding materials and labor. To meet his need, a multiple denaro coin, the grosso, was created. The design on the grosso showed a strong resemblance to Byzantine style, with a seated Christ on one side and the doge of Venice and St. Mark on the other. This common design was kept for more than 200 years. Often a small symbol would appear at the foot or by the side of the seated Christ figure (figures B57–B58).

Figure B57. Annulet by foot   Figure B58. Star by throne

Issues of Jacopo Tiepolo (1229–49), employed a pellet. Ranier Zeno (1253–68) issued grossos without any symbol or with a trefoil of pellets or an annulet. The annular solar eclipse of 1261 crossed southern Italy and may have the basis for the annulet. Lorenzo Tiepolo (1268–75) had a small triangle both at the side and at the foot, suggesting the appearance of a comet. In 1270, during the reign of Lorenzo Tiepolo, an annular eclipse passed directly over Venice (figure B59). Beginning with the grosso of Jacopo Contarini in 1275, and immobilized by various doges for the next 64 years, an annulet is found on the grosso, most likely as a representation of the annular eclipse.

Some numismatists have argued that these small symbols were mint marks. Metcalf (1979)

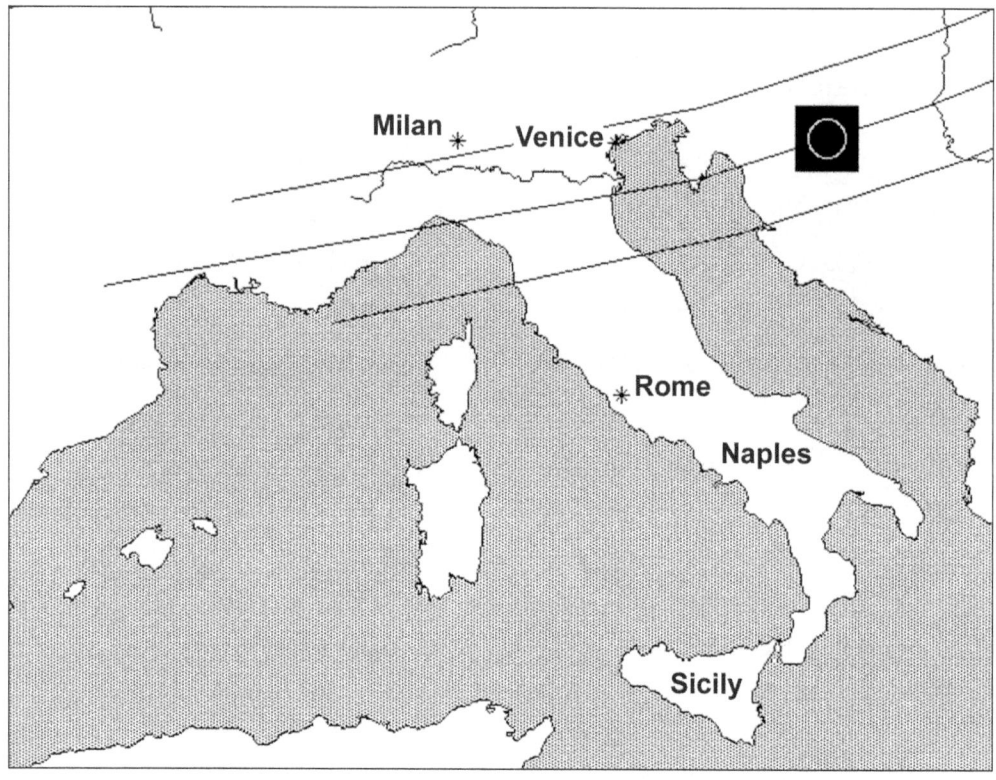

Figure B59. Path of the March 23, 1270, annular solar eclipse

makes a strong case that this conclusion is incorrect, but offers no alternative. He bases his rejection on hoard comparisons between Italian grossos and Serbian grossos of the same design. Astronomical symbols were used on Serbian pieces issued during the reign of Stefan Uros I (1243–76), which have a star between the feet of the seated Christ. The total solar eclipse of 1241 that crossed Serbia is a likely candidate for the star. The 1270 eclipse also crossed northern Serbia, and grossos of Stefan Dragutin (1276–82) and Stefan Uros II Milutin (1282–1321) contain a single pellet or an annulet at the base of the staff between the king and St. Mark (figure B60).

Beginning with the Venetian grosso of Andrea Contarini in 1368, a star was used in the design. This probably refers to the annular-total eclipse of 1366 that passed just north of Venice, although the eclipse did not become total until it was well to the east of Italy.

Figure B60. Annulet by staff

Castile

At three years of age, Alfonso VIII (1158–1214) inherited the throne. Mullets, stars, annulets, and crescents were used profusely in his coinage designs (figures B61–B64). These could refer to the total eclipse of 1178 and annular eclipses of 1153, 1192, 1201, and 1207.

One of the most interesting design types of Alfonso VIII portrays the king riding a horse on the reverse of the coin. Two variations can be analyzed in terms of astronomical events. Both

Figure B61. Stars and annulets

Figure B62. Crescent and mullet

Figure B63. Stars and king

Figure B64. Crescent

types have three pellets in a straight line as the beginning symbol on the obverse legend instead of the usual cross. This would imply that the pellets are significant. On the first type, the same three pellets are found beneath the horse on the reverse (figure B65). The second type replaces the pellets under the horse with a mullet (figure B66). As the pellets are on both pieces, one can assume that the mullet type was a design change that followed the first type.

Figure B65. Pellets under horse

Figure B66. Mullet under horse

The mullet on the second type may represent the total solar eclipse of 1178. If this is correct, then the three pellets may represent a three-planet conjunction that would have occurred between 1158 and 1178, probably one in which the planets were aligned. There are two candidate conjunctions to consider. On December 19, 1166, Mercury, Jupiter, and Saturn were in an extremely tight 0.7° triangular conjunction. While the closeness of the planets was extraordinary, their arrangement was not linear, and 12 years may have been too long to immobilize the coinage design.

A better choice might be the conjunction of November 30, 1176, when Mercury, Venus, and Jupiter were in a straight line separated by 4° (figure B67). During the preceding days, Venus and Jupiter had drawn increasingly closer together, and on the first day of December, the two brilliant planets, Venus and Jupiter, became one (separated by only 0.36°), with a thin crescent moon resting between these two planets and Mercury. Was this sighting taken as an omen for the reunification of Castile and Leon?

The only dinero of Henry I (1214–17) is without astronomical symbols. When he died the throne went to his sister Berenguela, who quickly abdicated in favor of her son Ferdinand, who was also the heir to Leon. Upon the death of his father in 1230, Castile and Leon were reunited.

During the reign of Ferdinand III (1217–52), no coins were struck with astronomical symbols. Of particular interest is the noticeable absence of any symbols marking the total solar eclipse that crossed Leon-Castile in 1239. This should have been recognized as a major event, as the last total eclipse to cross Leon-Castile was in 1079 and was probably used as the basis for a device on coins. The period 1230–50, however, was a time of systematic inquisition by the church against heretics. It may have been unwise to use astrological symbols in defiance of the pope, and Ferdinand was a zealous campaigner for Christianity.

Figure B67. November 30, 1176

The coinage of Alfonso X (1252–84) is also almost devoid of astronomical symbols. Perhaps this was a continuation of the response to pressures from the church. Alfonso X was born in 1225, and he should have observed the 1239 eclipse. He was known as "the Learned," and took counsel with rabbis in his investigation of astronomy and astrology (Lacroix 1878). One of his coins, however, is remarkable. On this dinero, which was issued near the end of his reign in 1284 (after the installation of Pope Martin IV), two clusters of five pellets are found below and at the foot of a lion (figure B68). The same five pellets are found not only by a lion on the one dinero of the infant prince, Henry (figure B69), but also by the sword on the pre-regal seal of Sancho IV (figure B70), who through intrigue and battle, secured the throne in 1284 that did not rightfully belong to him.

Figure B68. Late dinero of Alfonso X   Figure B69. Dinero of Prince Henry II

Figure B70. Pre-regal seal of Sancho IV

From June 1264 through December 1282 all five of the known planets came together 12 times in 30° to 40° conjunctions, but on December 12, 1284, the five planets were seen in the western sky just after sunset in a 12° cluster. Was this taken as a sign of Sancho IV's divine right to the throne?

Sancho IV (1284–95) issued two types of cornados, both with stars in the design (figures B71–B72). As he was not the legitimate heir to the throne, he may have used a variety of propaganda techniques to support his claim. While a total solar eclipse did pass just to the east of the Iberian peninsula, it occurred in 1295 at the end of his reign. In 1285, however, a total solar eclipse passed several hundred miles to the east of the peninsula and was seen as a 60 percent partial eclipse in Leon. Sancho may have stretched this event to his advantage, especially if the crescent types of Denis of Portugal (1279–1325) also referred to this event and Sancho wished to show dominance over the divine right of Denis. Although Portuguese knights fought beside Castilians against the Moors, they also had to continually defend their own borders against their Spanish neighbor.

Figure B71. Sancho IV cornado

Sancho was succeeded by Ferdinand IV (1295–1312), who began his rule at the age of nine years old under the regency of his mother. His coins, called noven, were of the standard castle on the obverse with a lion on the reverse. His first listed type, however, has three pellets in a vertical line on each side of the three-towered castle, with three small pellets on each side of the T (Toledo mint) below the castle, three annulets below the fore and hind legs of the lion, and one annulet above the lion's back (figure B73). This is certainly a significant addition to the standard coinage design. The use of three pellets and three annulets suggests that both types of symbols may represent the same event. The three pellets in a line suggest a linear three-planet conjunction as opposed to the triangular arrangement of the established cross and three pellets design.

Figure B72. Sancho IV

Figure B73. Noven of Ferdinand IV

On August 29, 1301, Mercury, Venus, and Saturn came together in a flat triangular fashion within a 1° circle. Even more significant, and closer to the beginning of his reign, was the conjunction of Venus, Mars, and Jupiter in the early morning skies of March 1298. The three planets approached each other in a straight line, and on March 16 were separated by less than 1°. The next day the brilliance of Venus merged with the brightness of Jupiter into a single magnificent object to the naked eye, with Mars next to them.

Sardinia

In Sardinia, the mullet remained as an immobilized design element until 1458, and was then sometimes used as a legend stop until 1598. The mullet was re-introduced with its full splendor by Charles II (1665–1700). On July 2, 1666, a total solar eclipse crossed just to the south of Sardinia. All of his crowns, first issued in 1671, portray the eclipse (figure B74).

Figure B74. Crown (10 reales) of Charles II

APPENDIX C.
SOURCES OF FIGURES

All of the figures except those listed below are drawn by the author. The numbers in **bold** are the figure numbers in this book. Some of the coin figures taken from other sources have been enhanced for clarity, corrected for omissions, modified to depict a new variety that has since been discovered, or augmented with additional coin illustrations by the author. Complete citations for the sources are shown in the bibliography.

Coin Figures

Baldwin (1915): **15**
Boudeau (1907–13): **153, 234**
Chutard (1871): **95, 148, 151, 152, 154**
Cohen (1880): **341**
Duplessy (1999), Jean Duplessy, les monnaies français royales, Paris 1999, with permission from the publisher, Maison Platt-Paris: **51, 128, 129, 273, 450, 461, 462, B39**
Folkes (c. 1850): **143, 144, 367, 369, 370, 373, 374, 375, 376, 378, 379, 383, 390, 393, 398, 399, 403, 407, 410, 422, 438, 437, 449**
Gariel (1883): **56, 208, 213, 214**
Hauberg (1974), with permission from Jørgen Sømod: **355, 356, 362**
Heiss (1865–69): **103, 105, 106, 118, 119, 120, 121, 122, 123, 124, 125, 180, 181, 264, 265, 266, 267, 269, 458, 459, A1, A4, B22, B23, B32, B33, B61, B62, B63, B64, B65, B66, B68, B69, B70, B71, B72, B73, B74**
Hoffman (1878): **127, 451, 452, 466, B36, B40**
Marcheville (1927–29): **B37**
Metcalf (1961), with permission from Spink & Son: **146, 147**
Poey d'Avant (1858–62): **21, 25, 50, 107, 113, 114, 126, 130, 131, 133, 134, 135, 136, 137, 159, 160, 162, 175, 177, 179, 184, 185, 186, 187, 188, 189, 190, 199, 200, 201, 202, 203, 204, 205, 206, 207, 230, 270, 272, 274, 276, 277, 423, 424, 425, 428, 429, 430, 433, 434, 435, 436, 439, 444, 445, 446, 447, 453, 454, 467, B3, B4, B5, B6, B7, B8, B14, B15, B41, B53, B54, B55, B56**
Réthy (1899–1907): **156, 223, 226, 227, 228, 283, B11, B19, B24**
Revue Numismatique (1920): **B38**

Robertson (1878): **B46, B47, B50**
Schlumberger (1878): **464**

Comet and Solar Halo Figures

Abell (1964): **166**
Guillemin (1875): **167, 168**
Mount Wilson and Palomar Observatories (1910): **164, 165**
Mount Wilson and Palomar Observatories (1957): **183**
Yeomans (1991), with permission from the European Southern Observatory: **176**

Ancient and Medieval Drawings and Figures

Bayeaux Tapestry (c.1066–76): **172**
Bondone (1303): **174**
Eadwine (12th century): **173**
English Psalter Map (13th century): **62**
Hevelius (1668): **233**
Historia Welforum (c.1185): **61**
Lacroix (1878): **Appendix A title page**
Langdon (1931): **27**
Lopez (1966): **Introduction title page**
Mausoleum of the Julii (3rd century): **347**
St. Paul's Epistles (c.1164): **63**
Sømod: **Appendix B title page**
Unknown source: **64**

BIBLIOGRAPHY

The term processed describes informally produced works that may not be readily available.

Abell, George. 1964. *Exploration of the Universe.* Austin, TX: Holt, Rinehart and Winston.
Ainslie, George Robert. 1830. *Illustrations of the Anglo-French Coinage.* London and Edinburgh: Hearne and Blackwood.
Alexander, Jonathan J.G. 1992. *Medieval Illuminators and Their Methods of Work.* London: Yale University Press.
Audouze, J., and Guy Israel. 1985. *Cambridge Atlas of Astronomy.* Cambridge, U.K.: Cambridge University Press.
Baldwin, Agnes. 1915. *Symbolism on Greek Coins.* New York: American Numismatic Society.
Barber, Malcolm. 1993. *The Two Cities, Medieval Europe, 1050–1320.* London: Routledge.
Bateson, J. D. 1997. *Coinage in Scotland.* London: Spink & Son.
Beeler, John. 1966. *Warfare in England, 1066–1189.* New York: Cambridge University Press.
Billings, Malcolm. 1990. *The Cross & The Crescent.* New York: Sterling.
Blackburn, Mark. 1991. "Coinage and Currency under Henry I: A Review." In Marjorie Chibnall, ed., *Anglo-Norman Studies XIII.* Woodbridge, Suffolk, U.K.: Boydell & Brewer.
_____. 1994. "Coinage and Currency." In Edmund King, ed., *The Anarchy of King Stephen's Reign.* Oxford, U.K.: Clarendon Press.
Boon, George C. 1986. *Welsh Hoards 1979–1981.* Cardiff: National Museum of Wales.
_____. 1988. *Coins of the Anarchy 1135–54.* London: Baldwin.
Boudeau, E. 1907–13. *Monnaies Françaises Provinciales,* 2 vols. Paris.
Brand, J. D. 1984. *Periodic Change of Type in the Anglo-Saxon and Norman Periods.* Rochester, U.K.: Self-published.
Brooke, G. C. 1916. *Catalogue of English Coins in the British Museum: The Norman King,* 2 vols. London: British Museum.
Bruun, Patrick M. 1966. *The Roman Imperial Coinage,* vol. 7, *Constantine and Licinius.* London: Spink & Son.
Bryant, Howard C., and Nelson Jarmie. 1974. "The Glory." *Scientific American,* September, pp. 60–71.
Carlyon-Britton, P. W. P. 1905–06. "A Numismatic History of the Reigns of William I and II (1066–1100)." *British Numismatic Journal* (British Numismatic Society, London) 2, 3.
Chutard, J. 1871. *Imitations des Monnaies au Type Esterlin.* Nancy, Belgium: Academy of Stanislas.
Cohen, H. 1880. *Description Historique des Monnaies Frappées sous l'Empire Romain,* vol. 7. Paris.
Cook, S. A., and others. 1961. *The Cambridge Ancient History,* vol. 12. Cambridge, U.K.: Cambridge University Press.
Cramer, Frederick. 1996. *Astrology in Roman Law and Politics.* Chicago: Ares.
de Marchéville, M. 1927–29. *Catalogue des monnaies françaises.* Paris.
DiMaio, M., J. Zeuge, and N. Zotov. 1988. "Ambiguitas Constantiniana: The Caeleste Signum Dei of Constantine the Great." *Byzantion* 58: 333–60.
Dolley, Michael. 1966. *The Norman Conquest and the English Coinage.* London: Spink & Son.
Duplessy, Jean. 1999. *Les Monnaies Françaises Royales,* vol. 1. Paris: Maison Platt.

Durant, Will. 1939. *The Life of Greece*. New York: Simon & Schuster.
Eggenberger, David. 1967. *A Dictionary of Battles*. New York: Crowell.
Elias, E. R. Duncan. 1984. *The Anglo-Gallic Coins*. London: Spink & Son.
Failmezger, Victor. 1995. "The First Vision of Constantine: An Alternative View of Campgate Coinage." *The Celator* 9(1): 20–24.
_____. 2002. *Roman Bronze Coins: From Paganism to Christianity 294–364 A.D.* Washington, D.C.: Ross & Perry.
Faintich, Marshall. 1995. *Symbolic Messengers of Medieval Man*. St. Louis, MO. Processed.
Fletcher, Michael. 2003. *Bloodfeud*. Oxford, U.K.: Oxford University Press.
Folkes, Martin. c. 1850. *English Coinage*. London.
Gariel, Ernest. 1883. *Les monnaies royales de France*, 3 vols. Strasbourg.
Gerald of Wales, c.1180–1220. *On the Instruction of Princes*. Translated by Joseph Stevenson, 1843. London: Seeleys.
Grant, Edward. 1994. *Planets, Stars, & Orbs; The Medieval Cosmos, 1200–1687*. Cambridge, U.K.: Cambridge University Press.
Grant, Michael. 1993. *Constantine the Great*. New York: Charles Scribner's Sons.
Grierson, Philip. 1991. *Coins of Medieval Europe*. London: Seaby.
Guillemin, A. 1875. *Les comètes*. Paris: Hachette.
Gurney, Gene. 1982. *Kingdoms of Europe*. New York: Crown.
Hallam, Elizabeth, ed. 1995. *The Plantagenet Chronicles*. New York: Crescent Books.
Haskins, Charles. 1927. *The Renaissance of the Twelfth Century*. Cambridge, MA: Harvard University Press.
Hauberg, P. 1974. *Danmarks mønter indtil 1241*. Frederiksberg, Denmark: Jørgen Sømod.
Hazzard, R. A. 1995. "Theos Epiphanes: Crisis and Response." *Harvard Theological Review* 88(4): 415–36.
Heiss, Alos. 1865–69. *Monedas Hispano-Cristianas*, 3 vols. Madrid: Milagro.
Hevelius, Johannes. 1668. *Cometographia*. Danzig.
Hewlett, Lionel. 1920. *Anglo-Gallic Coins*. London: Baldwin.
Hoffman, H. 1878. *Les monniaes royales de France*. Paris.
Hollister, C. Warren. 2001. *Henry I*. New Haven, CT: Yale University Press.
Jacob, Kenneth. 1985. *Coins and Christianity*. London: Seaby.
Johnson, Arthur Henry. 1877. *The Normans in Europe*. New York: Scribner.
Jones, A. H. M. 1986. *The Later Roman Empire*, 2 vols. Baltimore, MD: The Johns Hopkins University Press.
_____. 1994. *Constantine and the Conversion of Europe*. Toronto: University of Toronto Press.
Kieckhefer, Richard. 1989. *Magic in the Middle Ages*. Cambridge, U.K.: Cambridge University Press.
King, Edmund. 1994. "Introduction." In Edmund King, ed., *The Anarchy of King Stephen's Reign*. Oxford, U.K.: Clarendon Press.
Klawans, Zander. 1964. *An Outline of Ancient Greek Coins*. Racine, WI: Whitman.
Krupp, E. C. 1991. *Beyond the Blue Horizon*. New York: Oxford University Press.
Lacroix, P. 1878. *Science and Literature in the Middle Ages*. New York: Frederick Ungar.
Langdon, S. 1931. *The Mythology of All Races*, vol. 5, *Semitic*. Boston: Marshall Jones.
Lieu, S. N. C., and D. Montserrat. 1996. *From Constantine to Julian, Pagan and Byzantine Views*. London: Routledge.
Lopez, Robert. 1966. *The Birth of Europe*. London: M. Evans.
Mack, R. P. 1966. "Stephen and the Anarchy 1135–1154." In C. E. Blunt, H. H. King, and R. H. M. Dolley, eds., *The British Numismatic Journal 1966*. Dublin: Dublin University Press.
MacMullen, Ramsay. 1987. *Constantine*. London: Croom Helm.
Manchester, William. 1993. *A World Lit Only by Fire*. New York: Little, Brown & Company.
Markale, Jean. 1993. *The Celts: Uncovering the Mythic and Historic Origins of Western Culture*. Rochester, VT: Inner Traditions International.
Mattingly, H. 1967. *Christianity in the Roman Empire*. New York: W. W. Norton.
Meeus, J. 1991. *Astronomical Algorithms*. Richmond, VA: Willmann-Bell.
Metcalf, D. M. 1961. *The Coinage of South Germany in the Thirteenth Century*. London: Spink & Son.
_____, ed. 1977. "Coinage in Medieval Scotland (1100–1600): The Second Oxford Symposium on Coinage and Monetary History." *British Archaeological Reports 45*. Oxford, U.K.

_____. 1979. *Coinage in South-Eastern Europe 820–1396*. Special Publication no. 11. London: Royal Numismatic Society.
_____. 1988. *Yorkshire Numismatist* 1:13–26.
Molnar, M. 1997. "Mithrades Used Comets on Coins as a Propaganda Device." *The Celator* 11(6): 6–8.
Neugebauer, O., and H. B. Van Hoesen. 1959. *Greek Horoscopes*. Philadelphia, PA: American Philosophical Society.
Newton, R. 1972. *Medieval Chronicles and the Rotation of the Earth*. Baltimore, MD: The Johns Hopkins University Press.
_____. 1979. *The Moon's Acceleration and Its Physical Origins*, vol. 1, *As Deduced from Solar Eclipses*. Baltimore, MD: The Johns Hopkins University Press.
North, J. J. 1991. *English Hammered Coinage*, vol. 2, *Edward I to Charles II, 1272–1662*. London: Spink & Son.
_____. 1994. *English Hammered Coinage*, vol. 1, *Early Anglo-Saxon to Henry III c. 600–1272*. London: Spink & Son.
Olmstead, Denison. 1839. *Astronomy*. New York: Collins, Keese.
Pegge, S. 1772. *An Assemblage of Coins Fabricated by the Authority of the Archbishops of Canterbury*. 1975 reprint. Chicago: Obol International.
Pingre, P. 1783. *Cometographie ou Traite historique et Theorique des Cometes*. Paris.
Poey d'Avant, Faustin, 1858–62. *Monnaies féodales de France*, 3 vols. Paris.
Pradel, P. 1936. *Catalogue des jetons des princes et princesses de la Maison de France*. Paris: Bibliothèque National.
Rechenbach, Mary C. 1975. "The Gascon Money of Edward III." Ph.D. diss., University of Maryland, College Park, MD.
Réthy, Ladislaus, and G. Probszt. 1899–1907. *Corpus Nummorum Hungariae*. Budapest.
Revue Numismatique. 1920. Paris.
Richardson, Adam. 1901. *Catalogue of Scottish Coins in the National Museum Edinburgh*. London: Andrew.
Roberts, James N. 1996. *The Silver Coins of Medieval France*. South Salem, NY: Attic Books.
Roberts, W. R. 1895. *The Ancient Boeotians: Their Character and Culture, and Their Reputation*. Cambridge, U.K.
Roberts, W. R., and B. Head. 1881. *On the Chronological Sequence of the Coins of Boeotia*. London.
Robertson, J. D. 1878. *A Handbook to the Coinage of Scotland*. London.
Rostovtzeff, M. 1942. "Vexillum and Victory." *Journal of Roman Studies* (Society for the Promotion of Roman Studies, London): 92–106.
Runchman, Steven. 1958. *The Sicilian Vespers*. Cambridge, U.K.: Cambridge University Press.
Sauval, J. 1985. *A propos des cometes: orbites, luminosité, comètes remarquables*. In Observatoire Royal de Belgique Communications, Series A, no. 79.
Schlumberger, G. 1878. *Numismatique de l'Orient Latin*. Paris: Ernest Leroux.
Scott-Giles, C. W. 1929. *The Romance of Heraldry*. London: J. M. Dent.
Seaby, Peter. 1984. *Coins of Scotland, Ireland & the Islands*. London: Seaby.
Sear, David R. 1978–79. *Greek Coins and Their Values*, 2 vols. London: Seaby.
_____. 1987. *Byzantine Coins and Their Values*. London: Seaby.
_____. 1988. *Roman Coins and Their Values*. London: Seaby.
Steele, Joel Dorman. 1884. *New Descriptive Astronomy*. New York and Chicago: A. S. Barnes & Company.
Stephenson, F. R. 1997. *Historical Eclipses and Earth's Rotation*. Cambridge, U.K.: Cambridge University Press.
Stephenson, F. R., and L. V. Morrison. 1984. "Long-Term Changes in the Rotation of the Earth: 700 BC to AD 1980." *Philosophical Transactions of the Royal Society of London*, Series A, 313: 47–70.
Stevenson, Joseph. 1991. *A Medieval Chronicle of Scotland. The Chronicle of Melrose*. Reprinted translation. Felinfach, U.K.: Llanerch Press.
Sutherland, C. H. V. 1973. *English Coinage, 600–1900*. London: Batsford.
Tester, J. 1987. *A History of Western Astrology*, Wolfeboro, NH: Boydell & Brewer.
Ulansey, David. 1989. *The Origins of the Mithraic Mysteries*. New York: Oxford University Press.
U.S. Naval Observatory. 1991. *Astronomical Almanac*. Washington, D.C.: Government Printing Office.
von Oppolzer, Theodor Ritter. 1962. *Canon of Eclipses*. New York: Dover.
Warren, W. L. 1973. *Henry II*. Berkeley, CA: University of California Press.

Whitelock, Dorothy, David Douglas, and Susie Tucker, eds. 1961. *The Anglo Saxon Chronicle.* New Brunswick, NJ: Rutgers University Press.
Wyatt, Stanley P. 1966. *Principles of Astronomy.* Boston: Allyn and Bacon.
Yeomans, Donald. 1991. *Comets: A Chronological History.* New York: John Wiley & Sons.
Zimmerman, Linda. 1995. "Heads and Tales of Celestial Coins." *Sky & Telescope*, March, pp. 28–29.

INDEX

Aachen 53; *see also* Germany
Adhémar de Montil 15
Adrianopolis *see* battles
Aegina 10
Aethelred *see* England
Aethelred II *see* England
Aethelstan *see* England
Aethelwulf *see* England
Africa 38, 42, 100, 171, 175, 190
Ahenobarbus 190; *see also* Rome
Albigensian crusades 53; *see also* Feudal France
Alexander III *see* Scotland
Alexander Jannaeus 37
Alfonso I *see* Iberia
Alfonso II *see* Iberia
Alfonso IV *see* Iberia
Alfonso VI *see* Iberia
Alfonso VII *see* Iberia
Alfonso VIII *see* Iberia
Alfonso IX *see* Iberia
Alfonso X *see* Iberia
Alfonso XI *see* Iberia
Alfonso Henriques *see* Iberia
Alphonse de France 57, 155; *see also* Feudal France; Poitou
Amiens 23, 192; *see also* Feudal France
Amphipolis 36
Andreas I *see* Hungary
Andreas II *see* Hungary
Angevin 146, 149, 152; *see also* Anjou; England
Anglo-Gallic 50, 153–55, 157, 163; *see also* Aquitaine; England; Issoudun; Poitou; Ponthieu
Anglo-Saxon 13, 24–25, 65, 73–75, 78, 81, 95, 97, 123–26, 130–32, 140, 182; *see also* England
Anglo-Saxon Chronicle 65, 74–75, 97, 140, 182
Anjou 58, 92, 123–24, 140, 145, 150, 154, 171, 173, 175, 177; *see also* Anglo-Gallic; England; Feudal France
Anlaf Guthfrithsson 128; *see also* York
annular eclipse *see* eclipses
annulet *see* symbols

Antiochus I 83
Antiochus III 36, 62
Aphrodite 11, 15, 38
Apollo 30, 99, 101, 104
Apulia 12, 171, 193
Aquila 104
Aquitaine 50, 68, 71, 154, 160, 162–63, 165–66, 201, 206; Eleanor 50, 154, 162; *see also* Anglo-Gallic; England; Feudal France
Aragon 13, 42–44, 48–50, 171, 198, 208; *see also* Iberia
Arches 40; *see also* Feudal France
Armenia 14, 76, 78, 99, 103, 113, 198
Artemius 106
Artois 47, 192; *see also* Feudal France
Assize of Winchester 139, 141; *see also* England
astrology 3–4, 25, 31, 89, 212
astronomical symbols *see* symbols
Athena 9, 12
Athens 14–15, 37–38
Attica 38
Augustus Caesar *see* Rome
aurora 93, 173
Austria 45, 47, 53, 57, 156, 171, 197

Bamburgh 146; *see also* England; Scotland
Barcelona 48, 50; *see also* Iberia
Basil II *see* Byzantium
battles: Adrianopolis 103–4, 108; Chrysopolis 103–4; Hastings 63, 123, 131, 138; Magnesia 36; Marathon 9, 14, 116; Milvian Bridge 99, 102, 104–5, 107, 110, 113–14; Ourique 89; Philippi 190; Salamis 116; of the Standard 123, 146; Tinchebrai 123, 140
Bayeaux 63, 66, 133, 149
Bela IV *see* Hungary
Berry 72, 154, 159, 196
Blois 71, 123, 148
Boeotia 37–38
Bohemia 59

Boleslav II 59, 67
Boudicca *see* England
Bran 96; *see also* England
Brian fitz Count 149; *see also* England
Bristol 123, 146, 150; *see also* England
Britain 12, 100–1, 126; *see also* England
Brittany 44, 68, 71, 149, 159, 192; *see also* Feudal France
Brosse 72, 159; *see also* Feudal France
Brutus *see* Rome
Burgundy 51–52, 68, 71, 88, 91, 201, 203; *see also* Feudal France
Byzantium 13–14, 22, 25, 27, 45, 76, 94–95, 99, 103–4, 171, 173, 175, 179, 190, 193, 197–98, 207–9; Basil II 76; Constantine IX 22, 94; John I Tzimisces 76; John III 198; Michael VIII 173, 175

Caesar 61–62, 99, 101, 103, 111, 120, 190; *see also* Rome
Cahors 15; *see also* Feudal France
Calabria 36
Calais 72–73
Cambrai 25; *see also* Feudal France
Cambridge 113, 148; *see also* England
Canterbury 25, 29, 64, 74, 205; *see also* England
Caracalla 21
Caria 11
Carlisle 146, 206; *see also* England; Scotland
Carnuntum 101
Carthage 175; *see also* Crusades
Castile 13, 42–43, 48, 50, 88–90, 183, 200, 210–12; *see also* Iberia
Castor 18, 108, 110, 115, 117; *see also* Dioscuri; Gemini; Pollux
Catalonia 42, 48, 50, 171; *see also* Iberia
Celtic 12, 20–21, 28–29, 31, 35, 70, 85, 110; *see also* England

Index

Champagne 14, 51, 65–66, 92, 191; *see also* Feudal France
Charlemagne *see* France
Charles II de Gonzague 40
Charles IV 202
Charles de Valois 168
Charles of Anjou 58, 171–77
Charles the Bald 155
Chi-Rho 102–8, 110, 112–20; *see also* Christogram
Christ 26, 76, 97, 101, 105, 121, 198, 209–10
Christianity 4, 13–14, 22, 25, 29–31, 38, 99, 102–3, 105–6, 118, 121, 128, 174–75, 212 ; *see also* Trinity
Christogram 43, 98, 102–4, 106, 108–10, 112, 115–16, 118–19; *see also* Chi-Rho
Chronicles of Melrose 206; *see also* Scotland
Cilicia 14, 78, 198–99
Claudius Gothicus *see* Rome
Clodius 85
Cnut *see* England
Coenwulf *see* England
coma 60, 62, 65; *see also* comets
comb *see* symbols
comets 4, 18, 52, 59–80, 91–92, 103, 123, 126, 131–34, 136, 138–44, 146, 152, 165–66, 173, 178, 183, 192, 209; Donati 61; Halley 61, 63–64, 66, 68, 71, 74, 76–77, 92, 123, 131–33, 138, 146, 152, 165, 178, 192; Mrkos 59–60, 67; Seki-Lines 61
Commagene 83
conjunctions 5, 21, 22–24, 27–29, 69, 83–92, 104, 107–8, 110–14, 117–20, 125, 140–41, 159, 187, 205, 211, 213; *see also* planets
Conradin 171, 175; *see also* Charles of Anjou
Constans *see* Rome
Constantine *see* Rome
Constantine II *see* Rome
Constantine IX *see* Byzantium
Constantinople 28, 45, 64, 115, 118, 171, 173, 175; *see also* Byzantium; Rome
Constantius *see* Rome
Constantius II *see* Rome
Constantius Chlorus *see* Rome
Constantius Gallus *see* Rome
Constellations: Capricorn 21, 104; Gemini 18, 108, 111, 117, 120; Leo 83, 88, 90; Libra 23, 86; Orion 108; Pleiades 15–16, 18, 84, 86; Sagittarius 21, 104; Taurus 15, 82–84, 94, 96, 108, 117, 120; Ursa Major 18–19; Ursa Minor 19; Virgo 113, 158; *see also* zodiac
Corieltauvi 12; *see also* England
Corsica 171

Council of Clermont 173; *see also* crusades
Council of Nicaea 28, 121; *see also* Trinity
Crepy 92–93; *see also* Feudal France
crescent 9, 10–16, 18, 20–22, 29–30, 34, 37–38, 40, 44–45, 48–49, 51, 54–58, 68, 72–75, 80–81, 83–88, 90, 93–94, 106–9, 111, 113–16, 126–27, 129, 131, 142, 155–57, 162, 164–67, 169, 171–72, 183, 192, 195–96, 198, 200–1, 204–6, 210–13; *see also* eclipses; moon; sun
Crete 20, 36
Crispus *see* Rome
Croatia 32, 198
crusades 63, 136, 156, 170–77; Eighth 170–71, 175–77; First 173–74; Seventh 173–74
Cyzicus 114–15; *see also* Rome

Daedalus 20
Danube Bridge 99, 108; *see also* Rome
David I *see* Scotland
David II *see* Scotland
Delmatius *see* Rome
Delta-T 186
Demeter 20
Denis *see* Iberia
Denmark 54, 81, 126–30, 149, 189; Magnus the Good 127; Roskilde 126–27; Sven Estrithson 80–81, 130; Swein 78; Viborg 126–27; *see also* Cnut
Diocletian *see* Rome
Dioscuri 18, 108, 111, 115–17, 119; *see also* Castor; Gemini; Pollux
Domitia *see* Rome
Donzy 208–9; *see also* Feudal France
Dover 135; *see also* England
Dublin 128, 204; *see also* Ireland
Dumno Tigir Seno 28; *see also* England
Dura 21–22
Durham 136, 146, 148; *see also* England

Eadgar *see* England
Eadwine 64
Earth *see* planets
East Anglia 66, 74, 128; *see also* England
eclipses 11–15, 22, 32–58, 64, 68–69, 74–76, 80, 87–94, 126–31, 136–39, 142–44, 146–53, 155–62, 165–66, 168–69, 171–77, 179, 182–88, 189–213; annular 11–12, 15, 33–35, 37, 42–45, 47–53, 58, 68–69, 75, 80, 89–93, 127–28, 130–31, 136–39, 147, 150–51, 153, 155–62, 166, 168–69, 173–74, 184, 190–96, 198–206, 209–11; partial 11, 14–15, 22, 33–35, 37–39, 44, 87–88, 93, 127, 129, 136, 142, 156–60, 165–66, 174, 184–84, 190, 204–5, 213; total 12, 32–37, 39–42, 45–47, 53–58, 68–69, 74, 89–90, 93, 126–28, 142–44, 146–52, 162, 171, 174–76, 182–85, 187, 190, 192, 194, 197–208, 210–13
edict 101, 103
Edinburgh 146, 205–7; *see also* Scotland
Edmund 128
Edward *see* England
Edward I *see* England
Edward II *see* England
Edward III *see* England
Edward the Black Prince *see* England
Edward the Martyr *see* England
Egypt 62, 87, 101–2, 113, 174–75
Eighth Crusade *see* crusades
Elagabalus *see* Rome
Eleanor of Aquitaine *see* Aquitaine
Eleanor of Castille 154
Emesa 87, 94; *see also* Elagabalus; stone
England 4, 23–25, 42, 45–46, 50, 54–56, 63, 65–66, 68, 73–75, 78–81, 95–96, 101, 122–52, 154–69, 175, 190–91, 195–97, 203–8; Aethelred 74–75; Aethelred II 4, 78, 125–26; Aethelstan of Wessex 128; Aethelwulf 74–75; Berhtwulf 74; Boudicca 28–29; Channel 123, 134, 157, 191; Cnut 78–79, 126, 129, 131; Coenwulf 73; Cunobelin 31; Cuthred 73; Eadgar 129; Ecgberht 74–75; Edward I 68, 154, 157, 160, 163–69, 175, 207–8; Edward II 68, 153–54, 157–58, 160, 163–69; Edward III 68–69, 152, 154, 157, 160, 163–69, 206; Edward the Black Prince 154 (*see also* Anglo-Gallic); Edward the Confessor 79–80, 95–96, 123, 129–32; Edward the Martyr 125, 129; Harold 63, 123, 130–32; Harthacnut 127; Henry I 46, 97, 123, 125, 130, 137–45, 148, 152, 182; Henry II 50–51, 124, 145, 152, 154, 162, 205–6; Henry III 25, 54–56, 154, 162; John 154–56, 162, 195, 203–4; Matilda 123–24, 132, 146, 150; Richard (the Lionheart) 154–57, 162; Stephen 122–24, 138, 142, 144–52, 206; William I (the Conqueror) 63, 122–24, 130–38, 140, 150; William II (Rufus) 123–24, 130–39, 144
epsilon *see* symbols

Eudes 51, 75, 157, 207; *see also* Feudal France
Eudon 67–68; *see also* Feudal France
Eusebius 102–3, 105–6, 108, 112–13, 118; *see also* Rome

Fausta *see* Rome
Ferdinand II *see* Iberia
Ferdinand III *see* Iberia
Ferdinand IV *see* Iberia
Feudal France 14, 44, 47–48, 51–53, 58, 65, 67, 71–72, 78–80, 91–93, 154–55, 159–60, 166, 169, 191–92, 196, 203, 208–9
First Crusade *see* crusades
Flanders 23, 47, 57, 196, 208; *see also* Feudal France
Florence of Worcester 136, 182
France 23–24, 40–42, 45, 47–53, 64, 71–75, 78, 92, 96, 129, 152, 154–55, 158, 160, 162–63, 166, 168–69, 171–77, 190, 195, 197, 201–3, 207–9; Charlemagne 24, 73–74, 171; Jean II 202–3; Louis VI 22; Louis VII 22, 50–51, 92, 162; Louis VIII 173, 201; Louis IX 170–77; Louis the Pious 74; Philip II 22–24, 64, 78, 154; Philip IV 168–69; Philip VI 202–3
Franks 25
Frederick II 25
Frederick Barbarossa 25–26

Galeria Valeria *see* Rome
Galerius *see* Rome
Gallehus gold horns 189; *see also* Denmark
Gallienus *see* Rome
Gaston 40–41
Gaul 12, 36–37, 60, 100–1
Gemini *see* constellations; *see also* Castor; Dioscuri; Pollux
Geoffrey of Anjou 92, 123
Gerald of Wales 136
Germany 39, 45, 47, 53–56, 93, 123, 149, 171, 190, 195–97, 200; Brandenburg 53; Cologne 39–40, 207; Mainz 196; Regensburg 93; Swalenberg 54–56; Wismar 79–80; Wurzburg 196
Geza *see* Hungary
Giotto di Bondone 64
glory 97
God-given 22–24; *see also* Philip II
Greece 10–11, 15–16, 18–19, 23, 30, 33, 36–38, 113, 171, 186–87
griffin 30
Guthfrith 128; *see also* York

Hadad and Atargatis 21
Hadrian *see* Rome
Halley's comet *see* comet
halo 96–97

Hanniballianus *see* Rome
Harold *see* England
Harthacnut *see* England
Hastings 63, 123, 131, 135–36, 138; *see also* battles; England
Helena *see* Rome
Heliopolis 20–21
Helios 20, 22
Henry *see* England
Henry I *see* England
Henry II *see* England
Henry III *see* England
Henry VI 156
Henry of Anjou 124, 145, 150; *see also* Henry II
Henry of Blois 148
Henry the Lion 149
Hephaestion of Thebes 62–63
Heraclea 114, 116; *see also* Rome
Hercules 16, 100–1
Herod 29–30
Herodotus 33, 186
Hetoum 14, 76–77, 198–99
Hiberno-Norse 45–46, 69–70; *see also* Ireland
Hierapolis 21
Hoc Signo Victor Eris 99, 101, 119–20; *see also* Chi-Rho; Christianity; Constantine; Rome
Holland 45, 54–56, 190, 195
Holy Roman Empire 54–56, 171, 194
Hundred Years War 154
Hungary 45, 53–54, 57, 75–77, 93–96, 194, 196–98, 202, 207; Andreas I 94–96; Andreas II 196; Bela IV 57, 198; Geza 76–77, 194; Koloman 76–77; Stephen 75–77, 93–95, 194; Stephen II 77

Iberia 27, 42–44, 48–50, 88–91, 197–201, 210–13; Alfonso I 13, 44; Alfonso II 48–49, 77; Alfonso IV 200; Alfonso VI 43, 88–89; Alfonso VII 48–49, 89–90; Alfonso VIII 50, 183, 210–11; Alfonso IX 91; Alfonso X 212; Alfonso XI 200–1; Alfonso Henriques 88–89; Denis 200, 213; Ferdinand II 90; Ferdinand III 211; Ferdinand IV 200, 213; James I 198; James II 171, 208; Sancho I 89; Sancho II 77, 197; Sancho IV 212–13; Sancho VI 48–49; Sancho Ramirez 44
Iceni 28–29, 70, 85–86; *see also* England
Ionia 10
Ireland 25, 45–46, 126, 136, 152, 195, 203–4
Issoudun 51–52, 154, 156–57; *see also* Anglo-Gallic; England; Feudal France
Italy 35–36, 45, 47, 54, 73, 100–2, 171, 175, 197, 209–10

James I *see* Iberia
James II *see* Iberia
Jean II *see* France
John *see* England
John de Curcy 204; *see also* Ireland
Judah Aristobulus 37
Judea 29–30, 37, 87
Julia Mammaea *see* Rome
Julian II *see* Rome
Julius Caesar *see* Rome
Jupiter *see* planets

Kampanoi 11
Kaufbeuren 55–56
Knossos 20–30
Koloman *see* Hungary

L. Calpurnius Piso Frugi 30
Labarum 103, 112–18, 121; *see also* vexilla
Lactantius 102, 104–6; *see also* Rome
Lake Regillus 18
Langres 14, 91–92; *see also* Feudal France
Languedoc 15, 52–53, 79–80, 198; *see also* Feudal France
Leo *see* constellations
Leon 42–43, 48–50, 88–89, 200–1, 211–13; *see also* Iberia
Leon-Castile 42–43, 48–50, 88–89, 200–1, 211–13; *see also* Iberia
Leontius 27–28
Leopold of Austria 156
Libra *see* constellations
Licinius *see* Rome
Lincoln 123, 132; *see also* England
Lippe 54, 56
Lombards 25
London 39, 95–96, 110, 126–27, 129, 132, 184; *see also* England
Louis VI *see* France
Louis VII *see* France
Louis VIII *see* France
Louis IX *see* France
Louis the Pious *see* France
Lucretius Trio 84
Luna 85
Lund 81, 126–27, 130
Luxembourg 57
Lycia 13

Macedonia 11, 13, 36, 190
Magnentius *see* Rome
Magnus the Good *see* Denmark
Magyars 194
Majorca 67, 171–72
Malcom IV *see* Scotland
Man. Aquillius 84–85
Manfred 171, 175
Marcianopolis 87
Mark Antony *see* Rome
Mars *see* planet
Massilia 101; *see also* Constantine
Matilda *see* England

Maxentius *see* Rome
Maximianus *see* Rome
Maximinus Daia *see* Rome
Meaux 51–53; *see also* Feudal France
Mediterranean 35, 37, 171, 175, 207
Melos 15
Mercury *see* planets
Merope 18; *see also* Pleiades
Mesopotamia 15, 21, 30
meteorite 94; *see also* stone
Metz 16; *see also* Feudal France
Midlands 150, 191; *see also* England
Milan 31, 100, 103
Miletos 11
Milvian Bridge *see* battles
Minoan 30
minotaur 20
Mithradates the Great 62–63, 76
Mithraism 19, 96, 110–11
moneta nova 163–67; *see also* Aquitaine
moon 5, 7, 9–16, 18–21, 26, 29, 33–34, 37–38, 40–41, 45, 58, 60, 64, 83–87, 90, 93–94, 107–9, 111, 113–14, 116–17, 142, 158, 173–74, 184–85, 187, 196, 199, 204, 211–12; man in the moon 15–16, 204; *see also* planets
Moors 42–43, 88, 171, 175, 213; *see also* Iberia
Mount Athos 27
Mozarabs 43; *see also* Moors
mule 124, 132, 138, 144
mullet *see* symbols
Mysia 11

Naples 171
Navarre 48; *see also* Iberia
Nevers 39–41, 78, 208–9; *see also* Feudal France
Nicaea 28, 121, 198–99
Nicolas de Fontaines 25; *see also* Feudal France
Nicomedia 100, 104, 114, 116; *see also* Rome
Norman 25, 46, 95–97, 122–52, 154, 162, 165, 193; *see also* England
Normandy 123, 133–34, 140, 154; *see also* Feudal France
Norse 45–46, 69–70, 128
Northampton 135; *see also* England
Norway 126, 142
Norwich 78, 135, 150; *see also* England

Orion *see* constellations
Orleans 40–41, 92; *see also* Feudal France
Otho 130, 134

partial eclipse *see* eclipses
Patrick of Salisbury 148; *see also* England

Penthievre 67–68, 71–72; *see also* Feudal France
Persephone 20
Philip II *see* France
Philip IV *see* France
Philip VI *see* France
Philip Augustus *see* Philip II
Philip d'Alsace 23, 92–93; *see also* Feudal France
Phoenicia 16, 38
Picardy 192; *see also* Feudal France
planets 4–5, 17–31, 33, 35, 38–39, 69, 83–94, 104, 107–15, 117–20, 139–41, 179, 184–85, 205, 211–13; Earth 22, 31, 33–34, 39, 60, 74, 91, 94, 184–88; Jupiter 4, 17–19, 27–29, 69, 83–89, 91, 100–1, 104, 107–11, 114–15, 126–27, 140, 159–60, 205, 211–13; Mars 4, 18, 29, 69, 82–88, 90–93, 104–11, 116–17, 119–20, 140, 158, 205, 213; Mercury 4, 18, 27–29, 69, 83–88, 90–93, 107–9, 111, 114–15, 117, 125–27, 140, 198, 211–13; Saturn 4, 18, 27–28, 84–85, 87–88, 90–92, 104, 107–10, 117, 120, 125–27, 133, 140, 159–60, 198, 211–13; Venus 4, 18, 27–29, 69, 82–89, 90–94, 104, 107–11, 114–15, 117, 125–27, 140, 159–60, 198, 205, 211–13
Plegmund 75; *see also* England
Pleiades *see* constellations
Pliny the Elder 60–61, 79
Plutarch 38, 186
Poitou 154–56, 168, 195–96; *see also* Anglo-Gallic; England; Feudal France
Poland 47, 200
Pollux 18, 108, 110–11, 115–17; *see also* Castor; Dioscuri; Gemini
Ponthieu 154, 157–60, 166, 191–92; *see also* Anglo-Gallic; England; Feudal France
Pontus 11, 62
Porto 43; *see also* Iberia
Portugal 42–43, 53, 77, 88–89, 197, 200, 213; *see also* Iberia
Prince Henry 146; *see also* England; Scotland
Provence 48–49, 176–77; *see also* Feudal France
Psalter map 26
Ptolemy III 102–3
Ptolemy V 62
pyramid *see* symbols

Ralph, earl of Norfolk 133; *see also* England
Ravensburg 55–56
Raymond III 69, 165
Regulus 29
Reims 72–73, 191–92; *see also* Feudal France
reliquary 27

Rethel 39–41; *see also* Feudal France
Rhodes 36
Richard (the Lionheart) *see* England
Robert Bruce *see* Scotland
Robert of Gloucester 123–24; *see also* England
Roger II 175, 193
Roger, earl of Hereford 133; *see also* England
Roger of Hoveden 23; *see also* England
Rome 13, 17–22, 24–25, 28–30, 35–36, 39, 60–62, 82–88, 94, 98–120, 175, 190; Augustus Caesar 61–62; Brutus 190; Caracalla 21; Claudius Gothicus 101; Constans 99, 103, 115, 119; Constantine (the Great) 28, 98–121; Constantine II 99, 103, 113, 115, 118–19; Constantius II 99, 103, 113, 115, 119–20; Constantius Chlorus 100–1; Constantius Gallus 83, 120; Crispus 99, 102–3, 113; Delmatius 99, 103, 113, 115; Diocletian 100–1, 114; Domitia 19; Elagabalus 87–88, 94, 113; Fausta 103, 106; Galeria Valeria 114; Galerius 100–2, 114–15; Gallienus 20–22, 101; Hadrian 86–87; Hanniballianus 99, 103, 113; Helena 103, 105; Julia Mammaea 21; Julian II 82–84, 106; Julius Caesar 61–62, 190; Licinius 99, 101–4, 106, 108, 111, 113–15, 117; Magnentius 119; Mark Antony 190; Maxentius 99, 101–2, 105–6, 108–11, 113, 115; Maximianus 100–1, 106; Maximinus Daia 101; Severus 21, 101; Vespasian 35, 60; Vetranio 119–20
Romulus and Remus 115
rosette *see* symbols
Roussillon 92; *see also* Feudal France
Rudrasena III 22

Sagittarius *see* constellations
St. Ambrosius 31
St. Omer 47, 192; *see also* Feudal France
St. Paul's Epistles 26
Sancho I *see* Iberia
Sancho II *see* Iberia
Sancho IV *see* Iberia
Sancho VI *see* Iberia
Sancho Ramirez *see* Iberia
Sardinia 171–72, 213
Saturn *see* planets
Savary de Mauleon 156, 195–96
Saxa Rubra 102, 105; *see also* Milvian Bridge
Scandinavia 53
Scotland 45, 80, 123, 126, 130,

145–46, 150–52, 162, 191, 202, 204–7; Alexander III 206–7; David I 123, 145–46, 150–51, 191, 205–6; David II 206; Malcolm IV 204–5; Robert Bruce 206; William I (the Lion) 205–6
Seleucid 15–16
Sens 65–66, 76, 178; *see also* Feudal France
Seventh Crusade *see* crusades
Severus *see* Rome
Shrewsbury 132; *see also* England
Sicily 3–4, 58, 171–77, 193; *see also* Charles of Anjou
Sidon 16, 88
Sihtric 128; *see also* York
Simeon of Durham 136; *see also* England
Simon de Senlis 148; *see also* England
Sol 84–85, 99, 101, 105–6, 108, 110, 112, 116, 121
solar circles 94, 96–97, 125, 140–41, 144
Spain 4, 23, 42, 53, 100, 102, 182, 197, 200; *see also* Iberia
Spartans 37
Star of Bethlehem 29, 64
stars 11–15, 17–20, 32–33, 35–37, 39, 46–48, 50–57, 65–66, 68, 72–76, 79–81, 83–90, 94, 104, 106–15, 131, 133–38, 140–48, 150–52, 162, 165–66, 170–73, 175–77, 183, 190–91, 187–99, 202–3, 205–11; *see also* symbols
Stephen *see* England; Hungary
Stephen II *see* Hungary
stone 94; *see also* meteorite
sun 5, 7–12, 14–15, 18–20, 23, 26, 28, 31–58, 60, 83–84, 87–88, 94, 96–97, 99, 101–7, 111, 114, 121, 126, 136, 140, 142–43, 146, 155–56, 158, 170–77, 183–85, 187, 191, 195, 199, 204–5; *see also* planet
sun-dogs 96; *see also* sun
supernova 94–96, 125
Sweden 81, 126, 142
Switzerland 53, 171, 190
symbols: annulet 10–15, 25–26, 36–37, 44–46, 48–53, 58, 68–70, 79–81, 85–86, 89–93, 95, 113–14, 126–30, 132, 135–36, 139–41, 150, 155–60, 166, 168–69, 173–74, 193, 196, 200–3, 206, 209–10, 203; bar 51, 60–61, 66–68, 76–79, 157; comb 65–67, 69, 76, 91; complex 82–97; epsilon 67–71, 163–66; mullet 14, 35, 39–41, 54–58, 68–69, 90, 146–50, 155, 159–60, 165, 183, 198, 206–8, 210–11, 213; pellet 11–12, 18–25, 27–31, 35–36, 54–55, 70–73, 76, 78, 85–89, 92–93, 95, 110, 113, 116–21, 126–29, 133, 150–51, 173, 191, 196, 198–99, 203, 205–6, 209–10; pyramid 60–61, 70–76, 79–80, 92, 126, 131–33, 138–39, 144–45, 151–52, 192; rosette 69, 141–42, 145–48, 151, 159; swastika 13; three pellets 22–31, 55, 69–70, 85–86, 88, 91, 93, 110, 112–18, 121, 125–26, 159–60, 198, 210–11, 213; triangle 60–61, 70–77, 91–93, 209; triskeles 13; whorl 13; *see also* crescent; star
Syracuse 27
Syria 21, 62–63, 94, 101, 136, 173

Taras 36
Taurus *see* constellations
Tetrarchy 100–2; *see also* Rome
Thales of Miletus 33, 186–87
Thebes 37–38, 62–63
Theoderic 130
Thetford 136; *see also* England
Thibaut II 65; *see also* Feudal France
Thrace 12, 103
three pellets *see* symbols
Tiber River 102, 115; *see also* Rome
Toledo 13, 43–44, 89, 213; *see also* Iberia
total eclipse *see* eclipses
Treaty of Boves 23; *see also* France
Treaty of Falaise 205–6; *see also* England; Scotland
Treaty of Winchester 124, 152; *see also* England
Trebizond 27; *see also* Byzantium
Treveri 18–19; *see also* Rome
triangle *see* symbols
Trinity 22, 25, 27–28, 119–21; *see also* Christianity; three pellets
Trinovantes 70–71
Tunis 175–77; *see also* Eighth Crusade
Turkey 10, 33, 171, 186–87
Tuscany 24

Uranopolis 11–12, 18
Ursa Major *see* constellations
Ursa Minor *see* constellations

Varhran II 28
Venice 209–10
Venus *see* planet
Vespasian *see* Rome
Vetranio *see* Rome
vexilla 19, 21–22, 88, 113–16; *see also* Labarum
Viking 79, 128–29
Virgo *see* constellations
Visigoths 25
vision 6, 88–89, 98–12, 136; *see also* Constantine

Wallingford 149; *see also* England
Walram of Julich 39
Wareham 150; *see also* England
Watford 145–46, 148, 150–52; *see also* England
Western Satraps 22
whorl *see* symbols
William I (the Conqueror) *see* England
William I (the Lion) *see* Scotland
William II (Rufus) *see* England
William de Turnemire 163; *see also* Aquitaine
William, earl of Gloucester 150; *see also* England
William of Malmesbury 143, 146, 149, 182; *see also* England
Winchester 123–24, 133, 138–39, 141, 148, 152, 204; *see also* England
Wladislaw II 47

York 81, 101, 126, 128–29; *see also* England

Zeus 18, 62, 94
Zodiac 19, 110; *see also* constellations; stars

www.ingramcontent.com/pod-product-compliance
Ingram Content Group UK Ltd.
Pitfield, Milton Keynes, MK11 3LW, UK
UKHW050529150426
5217IPUK00026B/1868